"Enlivened by personal thoughts and examples this book serves up a refreshing blend of arts, science, anecdote and research, and provides useful stimulus in uncertain times to provoke thought and reconsider actions."

*— Professor Andrew Sharman CFIOSH FIIRSM FInstLM, author of* From Accidents to Zero, Safety Savvy *and* Mind Your Own Business

"This book presents a fresh perspective on the safety profession that is interesting, thought-provoking, and sometimes controversial. Approachable and relatable, yet profound and challenging."

*— Ron Gantt, Vice President and Principal Consultant at SCM Safety*

T0199294

# The Fearless World of Professional Safety in the 21ˢᵗ Century

Professional safety is in danger of extinction. Safety professionals have become complacent and unfocused, ignorantly relying on an 80-year-old paradigm. Lazy gimmicks are substituted for the hierarchy of controls meant to be the foundation of the profession. A $10,000 investment in posters makes zero improvement in safety; a $10,000 investment in machine guarding upgrades can save lives.

By blending philosophy, history, and psychology, *The Fearless World of Professional Safety in the 21ˢᵗ Century* is revolutionary, offering an innovative approach with creative solutions to move a safety program past the malarkey that has devalued professional safety for decades. Using humor and professional experience within a discussion of historical events and published scientific findings, Scott Gesinger explores the history of how current safety practices developed and why these must change if the profession is to survive the 21ˢᵗ century. He discusses new professional philosophies based on best practices in industry, historical examples, scientific research outside of safety, and proven approaches from other disciplines which can successfully guide safety professionals into the future.

Gesinger provides a book for every safety professional that is candid, plain-speaking, and eminently approachable, while at the same time provides information that is new, challenging, and engaging.

**Scott Gesinger**, CSP, is a safety engineer and writer in Minneapolis, Minnesota. He is a graduate of the University of Minnesota Duluth Master of Environmental, Health and Safety program and holds a Bachelor of Science in Psychology and Sociology from the University of Minnesota.

# The Fearless World of Professional Safety in the 21st Century

Scott Gesinger, CSP

Routledge
Taylor & Francis Group

LONDON AND NEW YORK

First published 2018
by Routledge
2 Park Square, Milton Park, Abingdon, Oxon OX14 4RN

and by Routledge
711 Third Avenue, New York, NY 10017

*Routledge is an imprint of the Taylor & Francis Group, an informa business*

*British Library Cataloguing-in-Publication Data*
A catalogue record for this book is available from the British Library

*Library of Congress Cataloging-in-Publication Data*
Names: Gesinger, Scott, author.Title: The fearless world of
professional safety in the 21st century / Scott Gesinger.
Description: Abingdon, Oxon ; New York, NY : Routledge, 2018. |
Includes bibliographical references and index.
Identifiers: LCCN 2017011734| ISBN 9781138036550 (hardback)
| ISBN 9781138036567 (pbk.) | ISBN 9781315178493 (ebook)
Subjects: LCSH: Industrial safety. | Safety engineers.
Classification: LCC T55 .G47 2018 | DDC 658.4/08--dc23
LC record available at https://lccn.loc.gov/2017011734

ISBN: 978-1-138-03655-0 (hbk)
ISBN: 978-1-138-03656-7 (pbk)
ISBN: 978-1-315-17849-3 (ebk)

Typeset in Bembo
by HWA Text and Data Management, London

# Contents

# Foreword

Something is happening in the world of safety. And it's big.

It is time, says Scott Gesinger, for a paradigm shift in safety. In fact, such a paradigm shift has already started. A fundamental shift in basic assumptions, in research approaches and professional practices is underway, if slowly and unevenly.

And *The Fearless World of Professional Safety in the 21ˢᵗ Century* is one of the books to help bring it along a bit farther still.

Professional safety has been perfused, for some eighty years now, by a particular idea about the role of the human in creating and breaking safety, and by a particular idea about how to contain that risk. In the old paradigm, people are a problem. They are unreliable, erratic; they get tired and complacent; their perceptual and cognitive apparatuses are limited and potentially misleading. The human is the problem; the worker is the weak link.

If people are a problem, then the road to safety is paved with efforts to contain that problem. For some eighty years, professional safety has done just that, or at least tried to. They have declared a 'war on error' to eradicate operator fickleness; promoted a 'zero vision' to ensure that even the smallest worker mistakes are declared undesirable and illegitimate. And of course, we have all seen a debilitating increase in the number of rules, procedures, checklists, and compliance requirements. Much work today is surrounded by such massive bureaucratic accountability requirements that it can no longer be argued that we simply have all that bureaucracy in place to deal with liability fears. There has been a rise in petty bureaucracy, in duplication of paperwork, reporting requirements, electronic surveillance, and more. What it has done is curtail human initiative, supposedly automate human unreliability away, shrink the bandwidth of worker innovation, initiative, and creativity, deny human autonomy and self-determination, eschew horizontal coordination of work in favor of top-down decrees for how to execute even the minutest tasks, and tighten supervision. We have been very successful in cluttering and clogging workplaces with safety stuff that does little but clutter and clog.

Of course, the distance between how work is imagined and how it is actually done, can only grow under the onslaught of such 'safety management.' And worse still, it is all stifling precisely the source of resilience that we need to deal with the residual safety challenges almost every industry still faces. Because, sure, work may never have been as safe as many think it is today. Yet at the same time, we have not made much progress on safety over the last twenty years. The number of fatalities in many safety-critical industries has stabilized, and however more bureaucracy and compliance we throw at the problem, we hardly seem to make a dent any longer. And the potential for system catastrophes and process accidents hasn't budged under our safety bureaucracy.

As with any budding paradigm shift, of course, there are a lot of people who'd like to interpret these data through the lens of the old paradigm. Remaining fatalities, for instance, are then seen as not enough compliance, as a lack of standardization, surveillance, and supervision. So that would call for even more of the prescriptions that come from the old paradigm. Tighter regulations, more control over the worker, more precise prescriptive paperwork. Workers need to be told to try harder, to follow procedures more carefully, to watch out better. This book calls on us to abandon some of the cherished islands of security the safety profession has always retreated to when faced with apparently insurmountable problems: standardize more, shrink the operational bandwidth for human action, turn the worker into the problem and object for intervention all at the same time, exercise more centralized control, try to change worker behavior by targeting worker behavior rather than understanding and targeting the circumstances, contradictions, and complexities under which they work.

This book, consistent with others that push the paradigm and that want to prepare the ground for a new one, calls for something else altogether. It calls for something fearless, rather than more of the same. The world of professional safety in the 21st century will require professionals who are willing to face the data, to acknowledge the facts. People don't get killed by walking around with a coffee cup without a lid. An oil rig doesn't blow up because a supervisor didn't hold a handrail. If we want to make progress on safety in industries where progress has stalled, the only place we can still go, is one where people have more autonomy, not less; one where we get together to push back on compliance and bureaucracy; one where we are open to other ways of working, to understanding the complex sources of both failure and success. And it's a place that requires fearless safety professionals. Welcome to the 21st century!

**Professor Sidney Dekker**
Director, Safety Science Innovation Lab
Griffith University, Brisbane, Australia

# Introduction

## It's the end of the world as we know it (and I feel fine)

*Herbert William Heinrich*
Heinrich worked for Travelers Insurance Company about a century ago. His contributions include "Heinrich's Law," which says that for every 300 near misses or minor incidents, there will be one major injury (like a fatality or permanently life-altering event). He also declared that most workplace injuries are due to worker behaviors (Heinrich, 1931). I will discuss some more of Heinrich's concepts in detail throughout the book.

*Fearless* is meant to accomplish two goals. First, it is meant to introduce "outsiders" to the world of professional safety. Outsiders are the people we safety professionals work for, and many of you who may at some point venture bravely into the world of professional safety. Most of the sidenote sections you will see are meant to explain concepts that may be basic information for more experienced safety nerds. Goal number two is to discuss how professional safety must change to thrive in the brave new world of the 21st century.

*Safety Nerds*
This term is a compliment. Nerds rule the world (the sticker on my Nintendo Wii says so!) and without the academic and philosophical dedication of people who show an interest above and beyond that of the average person to the art and science of safety we would live in a much more dangerous and chaotic world.

This book takes the key concept of modernizing the art and science of safety, and mixes it with experiences I have had over my decade and a half as a safety professional. The personal stories I share are meant to make the reading more relatable (possibly even entertaining, which seems to be a cardinal sin within safety), and to tie concepts to real-world situations. Very few of my experiences are unique to my career, and my hope is that readers will gain a sense of vocational kinship. We safety nerds oftentimes feel like islands in a professional sense, and it is always nice to know that someone else has gone through some similar situations. This book is also meant to be a bit light-hearted. I will use terms like "safety nerd" in a jesting sense; so much of what safety professionals do is serious that it is nice to smile and chuckle now and then about our jobs. For those of you outside of the safety profession who might be reading this, one of my hopes is that you will gain a better understanding of the folks from the EHS office and why we're so jittery all the time.

I want to deliberately challenge some common philosophical notions within the safety profession. I will be very upfront about how and why the concepts I've witnessed and worked directly with have succeeded or failed over the years. Your experiences may have resulted in different outcomes. While no offense is meant to any party who strongly believes in a certain concept (or pays tons of money for it), I do want this book to be like a nice splash of cold water; it might startle you a bit. The most important thing to remember as you read about the experiences I share is that the success of various styles and tactics depends first and foremost on sincerity. No matter what you try, *you* have to always just be "you."

It is time for a paradigm shift in safety. As practitioners, safety professionals need to relearn our art and science to reflect the soundly researched and tested concepts that have begun to take shape in the 21$^{st}$ century. For example, Heinrich's Law is nearly a century old and was based on data that was collected in a manner most modern researchers would reject (Manuele, 2002). Academic programs dedicating time to teach safety students about Heinrich as if his concepts are still valid is akin to an astronomy program dedicating time to teaching students Aristarchus of Samos's model of the earth residing at the center of the universe. Researchers like Fred Manuele are safety's version of Copernicus, and the way we approach our profession should change accordingly as new knowledge is gained.

Likewise, the practical approach to administrating safety programs must shift along with our concepts. Posters, slogans, games, and other efforts focused on individual employees siphon away precious time and monetary resources that could be dedicated to identifying and solving the root issues within the overall management system affecting the safety of an organization. While the older concepts mentioned above have been important stepping stones on the path to our modern view of safety and Heinrich was essential in the evolution of the profession, it is time for the old ideas to be swept into history and for safety practitioners to seek out the next concepts. This is a natural progression for any science; if computer engineers never changed, I would be typing this book with an old Remington typewriter and programming my work computer with a series of hole-punched cards.

The world is on the cusp of incredible change. Technological advances will present us, in the very near future, with a plethora of new opportunities for the advancement of safety for both humans and the environment. Self-driving cars will eliminate or greatly reduce driver error, drunk driving, aggressive driving, police chases, distracted driving, unsafe speed, and an untold number of other hazards. How will this affect the safety profession? Hanging a safe driving poster in the employee lounge of a driver dispatch office becomes useless when the truck driver is more of a GPS operator and cargo caretaker than an operator of a motor vehicle. Putting together hours' worth of educational material regarding the hazards of drinking and driving becomes a thing of the past the day anyone can stumble into their car and be ferried home safely by an autonomous vehicle, although this also means that other people will be exposed to bodily fluids when cleaning the insides of cars after a partier has messed it up with vomit or… well, you can use your imagination to guess what might end up in there when drunk people decide to share a ride home. 3D printing will reduce the need to move and handle heavy molds, robots are learning to use artificial intelligence to avoid contacting humans in their work space, and smartphone apps connected remotely to a sampling device will soon be able to monitor air and water discharges with terrific accuracy. To put it plainly, technology has the potential to make an ultra-safe and clean workplace the rule rather than the exception. Our challenge as safety professionals is to keep up, adapt, and remain relevant. "Impossible" simply means we haven't figured out how to do something yet. In 1919, it was impossible to get from Minneapolis to Sydney, Australia in a day. Now, it takes under twenty hours; because we figured it out. Our challenge as safety professionals is to keep up and adapt as the impossible becomes the everyday.

The role of the safety professional will change with these changes in technology. To stay viable, we have to shift focus so that we are aware of the available technology and can think about creative and innovative ways to use existing and emerging technology to provide safety management practices that not only make work safer, but smarter and more efficient. In a matter of years, we'll have robots as useful as Rosie from *The Jetsons*. It is up to you, safety professional, to point out that if she can vacuum the carpet and clean the windows while making wisecrack comments, then she can probably enter a permit-required confined space and do some work that would otherwise be dangerous for a human.

Good luck out there!

# 1 Watch *Jaws*

"Watch *Jaws*" are the two words I use to respond to people who ask me what it is like to be a safety professional. The film *Jaws* is an almost perfect model of the profession I've chosen. The book is excellent, too; but the movie is really a better model for the challenges of professional safety. Warning: There are spoilers ahead! If you've never watched *Jaws* or haven't seen it in years, immediately put down this book and go watch it! I'll wait right here. Dum dee doo dee dum…

Okay, you're back. Here we go! Chief Brody is new to town and, more importantly, new to the *culture* of the town; just as a freshly hired safety professional is new to a factory or other work setting and its culture. There are scenes early in the movie where Brody is acclimating to the town, learning their accents and getting an earful from a business owner whose fence has been "karate chopped" by some local youngsters, just as a safety professional has to learn the company jargon and figure out who the Activists (more on them when we cover safety committees) in the company are.

When the first victim is found, it significantly shakes up everyone involved, and the medical examiner determines that the young woman died of a shark attack. Brody responds by closing the beaches; he's never dealt with a shark attack before and isn't sure how to proceed. In the meantime, he learns that there is a group of Boy Scouts performing their one mile swim, so he rushes out to make sure everything is fine. On the ferry, he gets cornered by the mayor and other city leaders, who intimidate Brody into keeping the beaches open; the medical examiner tells Brody that the earlier death wasn't a shark attack but a boating accident. This mirrors a safety professional being intimidated by a management team that is more concerned about production numbers and profit, or key performance indicators centered on incident rates. Most of us in the safety profession can relate well.

After Brody grudgingly agrees to keep the beaches open, it is obvious that he isn't comfortable with the decision. It isn't long before the shark kills again, this time a young boy. When the boy's mother finds out that Brody could have closed the beaches but didn't, she slaps him hard across the face and chews him out in front of the entire town. The mayor, who had bullied Brody into keeping the beaches open, tells Brody the woman was wrong and Brody replies simply, "No she wasn't." Here we've got the safety professional coming to terms in the most difficult way with his responsibility. He knew what he should have done to keep people safe, allowed himself to be pushed into not doing it, and there was another incident. Brody is devastated.

---

*The BLS*

The Bureau of Labor Statistics collects data about American workplaces. This data is available for public consumption at www.bls.gov. The BLS website includes injury and illness data, unemployment data, inflation data, and lots of other really useful information. Check it out; you paid for it.

---

During this same scene, there are crowds of fishermen trying to push each other out of the way as they race to find the shark—and get their hands on the $3,000 reward ($13,306.96 in 2015 dollars according to the Bureau of Labor Statistics) the dead boy's mother is offering to the person or group that kills it. Brody is pleading with them not to overload their boats or perform unsafe acts on the water, but they don't listen. Here's a great example of the safety professional who has no control over the culture, and can see more accidents on the horizon.

Matt Hooper, a shark expert, arrives in town to help out. He immediately recognizes that a shark some fishermen have caught isn't the one that is responsible for the attacks, but the only person in town who will listen is Brody. More than once I've brought consultants in who see exactly what I see the way I see it but get zero traction with management. Like Hooper, they usually want to leave the situation quickly, but unlike many consultants, Hooper decides to stay. I think the reason he stays is that the fundamental difference between Hooper and a safety consultant is that Hooper is working pro-bono for the greater good.

When Hooper examines the remains of the first victim, he has stern words for Brody and the medical examiner. It takes Hooper about two seconds to see that the woman had obviously been killed by a very large shark, and he is understandably upset. Brody, like many safety professionals, has to admit to himself that he should have stood his ground; we never see this realization verbalized in the film, but it is obvious that he understands what he should have done, much like many safety professionals who have been called to task by an outside set of eyes (like an OSHA inspector or a consultant) after they've caved to management.

---

*Citing an OSHA standard*

OSHA standards are written in a Code of Federal Regulations, or CFR. The General Industry standards are 29 CFR. Standards are further broken down by paragraph using an outline style. Therefore, an example of a typical citation of an OSHA standard is 29 CFR 1910.144(a)(1)(ii). Go ahead and look that one up on the OSHA website, it is good practice.

---

That evening, Brody's son is sitting in his birthday gift, a sailboat that is tied to a dock outside the family's beachfront home. When Brody finds out, he yells at his son to get out of the boat but Brody's wife tells him to relax; that is, until she sees a picture in a book of a shark attacking a boat. At that point, she begins yelling at their son as well, but with even more zeal than her husband. Safety professionals go through similar situations frequently, and oftentimes not even our spouses can fully relate to our struggles until they somehow get a taste of the experience itself. Later, when Hooper stops by, Brody hasn't touched his dinner. A bottle of wine is opened, and Brody gets a little tipsy. Also not uncommon among safety professionals. One of my many mantras is "Work hard, play hard," and as long as it doesn't become a problem, I am fully on board with a few drinks to blow off some steam after a stressful day or week. When I go to conferences, it is not uncommon for me to spend a night or two out late with other safety nerds having drinks and telling war stories with comrades who "get it."

Brody and Hooper come across another attack that night as they cruise the local waters looking for the shark. The next morning, they ask the mayor again to close the beaches. The mayor won't give Hooper and Brody the time of day. The mayor tells Brody to do whatever he needs to in order to keep the beaches safe, but the beaches cannot be closed. The best thing to do to keep the beaches safe is to close them, but instead, resources are expended on extra boat patrols, helicopters, and other efforts that completely ignore the root cause of the problem. In safety, how many times does this happen? Again, most safety professionals can relate to experiences in their own career.

There is another attack, this time in front of hordes of beachgoers. Brody's son is hospitalized with shock, having seen the giant shark swim right past him as it was eating another boater. The mayor is in almost as much shock as Brody's son, and finally agrees to close the beaches and pay a local expert fisherman $10,000 ($44,356.51 in 2015 according to the Bureau of Labor Statistics) to find and kill the beast. Here we have a reckoning that many production managers come to only when they experience one of their employees get hurt badly or killed. The realization that safety actually does matter, that bad accidents can happen to their people, and that they will feel responsible for it for the rest of their life. I've thought in the past that all new managers should somehow go through what it feels like to experience an employee fatality, because once they do, almost all of them have a completely different outlook about safety. Unfortunately, I don't know of any way to adequately simulate the experience.

The rest of the movie is the hunt for the shark, and has fewer parallels to the safety profession. It is interesting to see how Brody learns about the boat, fishing, sharking, etc. in the way that many safety professionals realize just how green they are when they take a deep dive (pun intended) into the activities employees are performing.

Throughout the movie, Brody struggles with the needs and wants of the town. He has innumerable people pressuring him to keep the beaches open, because the livelihoods of the population are dependent on summer tourists. He struggles to balance the economic needs of the town against the safety concerns regarding the shark. It is a series of no-win forced decisions for the Chief; if he closes the beaches, most of the town spends the rest of the year on welfare. If he keeps the beaches open, people could die. Either way, he holds an immense responsibility on his shoulders that is likely to result in negative outcomes and alienated "customers." This is our curse as safety professionals, our burden, and our sacred responsibility. We *must* make the decisions that nobody else will… that nobody else *can*.

The ultimate question is, given all this pressure and the repeating sets of poor options to choose from, how does a safety professional avoid burning out? It isn't easy, but the following are some things individuals can try, mostly focused on living with the decisions they've made.

My primary personal goal is always to pass what I call the "Sleep Well Tonight" test. I know that I'm going to have challenges that parallel those of Chief Brody in *Jaws*, so when I am struggling over these choices I place the most weight on which one will allow me to sleep better that night. In other words, if I know that Option A will keep me awake and staring at the ceiling more than Option B, I choose Option B. Oftentimes when I discuss an issue with management or employees and they point out that I am taking them beyond an OSHA standard, or that there is no OSHA standard that addresses the issue at all, I tell them "I don't care what OSHA says about it, I have to sleep tonight." This strategy has worked out well for me since I started using it and it is hard for others to argue against it. Even if people don't understand the pressures of my job, they do understand the need for an individual to feel comfortable with the decisions they make.

The next goal is to pass the "60 Minutes" test. I learned this one in my college years when I worked at a department store as a loss prevention officer, arresting shoplifters. Throughout the day, with almost all the professional choices I make, I think about whether or not I would make the choice differently if a camera crew from *60 Minutes* was following me, filming an episode about being a safety professional. If I would not choose Option A if I were on camera, why would I choose it with the anonymity I have at work? Using this test forces me to examine my choices from a big picture perspective; many times it has helped me stand up to groups of people pressuring me to make a choice I am not comfortable with. Other times it has helped me have enough patience to take some extra time to really turn over decision options mentally to see them from every angle. Ultimately, this test helps guide me to make morality a prime factor in how I conduct my business.

My next strategy is one that is easy to share with managers, employees, or others who want to set safety aside in order to hit production numbers, profit goals, etc. I think about, or ask the others involved to think about, whether or not they would want their wife, parent, brother, best friend, or *child* to perform the work under the conditions we are evaluating. Have an employee who keeps taking the guard off a machine? Ask him if he'd be OK with one of the people previously named doing the same thing. Have a manager who refuses to install railings or fall protection tie-off points? Ask that manager if they would want their loved ones to do that job without the proposed protections. It can be a game-changing question, because it forces people to think about the value of their loved ones balanced against the production concerns they are pressuring you with.

Most safety professionals I know have diversions to allow them to mentally and physically "decompress" as they need to. I write books (some of them quite silly), exercise regularly, and turn into a little boy every time I play with my son. A good friend and colleague of mine more or less produces his own music, another cooks, and another I know runs daily. All of these safety professionals face the same struggles that I do, that Chief Brody did in *Jaws*, and that you probably do, too. What is your diversion? Is it healthy?

For students looking to enter the field or established professionals who want to make a career shift into safety, it is important to understand that the stresses you are about to face are unlike anything you've dealt with before. As I will state again later in this book, if you do this job long enough you will have to deal with a horrible injury or a fatality. Likely, more than one over the course of a career. Every day, you are going to be faced with hard choices, and there will be times that your phone rings and you get a cold chill, irrationally fearing what the caller is contacting you for. This is not a profession for the weak-willed, the fragile hearted, the thin-skinned, or the easily unnerved. People who have witnessed my day-to-day work and commented they would never do my job include prison guards, fire chiefs, law enforcement officials, and members of a company's senior leadership team.

> *Show me the money!*
> According to the Board of Certified Safety Professionals, the median salary for Certified Safety Professional (CSP) in the USA in 2015 was $109,000. To become a CSP, one must have a number of years of experience and education, then pass some very difficult exams. I tell people that a CSP is like a Jedi knight of safety.

Hopefully, there are some people who are not safety professionals that are reading this book in an effort to relate to a safety nerd. You might be the spouse of a safety professional (so sorry about that, my wife feels for you), the boss of one, a member of management, the parent of a safety nerd, or somebody who thought this was a Michael Crichton novel and didn't find out until you were on the airplane and couldn't return it that you grabbed the wrong item off the shelf. How does this chapter apply to you? The challenge for you is to relate to us. Understand why we do what we do, and why we do it the way we do. Understand that sometimes it feels like our career is about getting kicked in the crotch for eight hours a day, five days a week. Safety professionals aren't in it for the money, even though the paychecks are usually not bad. We aren't in it for fame, glory, an easy day at the office, or because we enjoy visiting the hospital to talk to someone who's just had his leg amputated. We do this job because of a deep-rooted desire to improve the human condition.

We are the people responsible for everything from that little strip of metal that keeps the carpet from pulling up where it meets the linoleum in your kitchen to the multiple redundant systems on the aircraft that flies you to your vacation destination. The next time you look at a playground that has been redesigned to meet safety standards, have to take seven guards off of a machine that you need to perform maintenance work on, or get a ticket for not wearing your seat belt, remember that *every* safety rule is written in blood. Children died and were horribly injured before anyone would agree to strict playground safety standards. Workers died, had body parts amputated, and were maimed for life before people decided we needed machine guarding standards. Countless men, women, and children were killed in auto wrecks from injuries that were completely preventable before laws were passed requiring seat belt, child car seat, and booster seat use.

Practicing safety professionals and the safety professionals who came before us, whether you see us as valued assets or dreaded safety cops, may have already saved your life a dozen times over without anyone ever knowing it. It may be that the plane you flew on didn't crash because of a proper inspection checklist, the brakes on your car were effective enough to stop in time because of the standard requiring brake lines to be built the way they are, the wiring in your house not catching fire because the electrical code was in place, or any one of a billion other non-events that have kept you with us this long. You are welcome, we're happy to help.

# 2 Something has to change

If you read the introduction, you've figured out that a clear message of this book is that it is time for the safety profession to shift into new ways of thinking for the 21$^{st}$ century. This became clear to me while attending a recent professional development conference. It was the second day of the conference, and I was meeting with some peers after the sessions had ended. We compared notes, and although we had all attended different sessions, most of us could accurately guess what the main points of the others' sessions had been. The regular buzz words floated back and forth between us, words like "transformational change." The presenters talked about the value of safety, but few of them itemized the skills, abilities, or specific practices that are part of that value. There were no deep dives into engineering economics, or examinations of how to problem-solve at a line-employee level. Safety philosophy had, we concluded, grown stale. Even worse, many of the sessions were presented by people who wanted to sell us something.

"I'm tired of hearing the same stuff over and over. I'm going to write a book about safety!" I declared. One of my peers (a friend and former coworker) rolled his eyes at me and left to get a beer.

This exchange is at the heart of where our profession stands. Safety nerds collectively seem to feel like we are living the infamous and inaccurately credited quote "Everything that can be invented has been invented." That quote, by the way, was attributed to Charles Holland Duell, Commissioner of the United States Patent and Trademark Office from 1898 to 1901. He never actually said it (Sass, 1989). Something Duell *did* actually say in 1902 should be the way safety professionals look at the future: "In my opinion, all previous advances in the various lines of invention will appear totally insignificant when compared with those which the present century will witness. I almost wish that I might live my life over again to see the wonders which are at the threshold" (Chances for the Inventor, 1902).

Bam! Duell nailed it! He knew that technology was far from stale, far from over, and far behind the level of advancement it would see in the next century. His statement, a bright and optimistic look to the future of technology, came before the Wright Brothers mastered controlled heavier-than-air flight, before digital computers, Bakelite plastic, television, or Robert Goddard's liquid-fuel rocket were all patented. If Duell had lived his life again, starting the day after he died, January 20, 1920, he would have lived to see the internet, nuclear power plants, cell phones, moon landers, and thousands of other wonders.

In safety, we stand in a place similar to that of Duell when he looked to the future and wished he could live another lifetime, if only to see the wonders that would be invented. Nobody can definitively know what is yet to come that will revolutionize safety. Our Wright Brothers might be two siblings who are learning to create apps right now; the next big step in computing could be integrating digital tech into human biology to a further degree; nuclear fusion may create limitless free power; new nanomaterials will make old style plastics look ridiculous and antiquated to my children the way thatched roofs look to me; as televisions fade away, employees are spending more time distracted by the entertainment and socialization value of their phones; and before I die of old age (no other way to go, right?) I will probably witness a human walk on Mars. Holy buckets, what an exciting time to be alive!

Our key to the future as safety professionals lies in the willingness to not just think outside the box, but to crush the box and throw it in the recycling bin. As new generations enter the workforce with new ideas and attitudes, the ways we have practiced our profession in the past will become less relevant. One simple example is the way in which people attend meetings.

---

*Lockout*
Lockout, also known as lockout/tagout, is a method of placing a locked barrier on a machine to prevent the unexpected release of energy. The most simple example is unplugging a device, placing a canister over the plug, then locking that canister closed.

---

I recently found myself standing in front of a group of employees who were about to attend annual lockout training. When I say they were about to attend lockout training, what I really mean is that they were about to be told the same information about lockout they'd heard for the last five years; of course, I always change the presentation and never rely on videos to do my job, but how many different ways can you really deliver the same required information every year?

They all had their phones laying on the tables in front of them. One even had his laptop computer open. I decided to call an audible (a football term for when the quarterback changes the play at the line of scrimmage). I only touched briefly on the information in my PowerPoint slides, and with each chunk of information, challenged the group to find something related to that information online using their phones.

Lockout hardware has to be of a standard color and construction... Find me three options online for a ball valve lockout device that meet this requirement. Our company requires locks to be individualized... Find three vendors who offer various ways for us to do that. Then we got to a trivia break (I like to throw in a safety trivia question every few slides to keep things from getting too boring). After we answered each trivia question throughout the presentation, I asked people to find a video online related to the question. Training went long, but the feedback was remarkably positive.

It is this type of shift that tons of other safety professionals are making even more effectively than I can, that marks a change in what we do and how we do it. The primary change, however, has to be centered on our attitudes toward safety. I fully agree that safety works best using a top-down strategy, although I am really curious about a new model I've lately heard a few people talk about that says safety works best when thought of like a sandwich, using a top-down *and* bottom-up approach at the same time.

---

*EHS and OHS*
The term "EHS" means environmental, health, and safety. Some people put it into another order, and say SHE. Tomato, tomato. Wait, that doesn't work in print. Oh, well. A typical safety professional also handles environmental compliance, which gives us the "E" in EHS. The "H" primarily involves evaluating chemical exposures, but as you'll read later in the book, I think it should involve more than that.

Some people have begun using the term "OHS," where the "O" stands for "occupational."

---

I haven't learned enough about the sandwich model to expound on it here, but the fact that it is a new and novel way to approach safety intrigues me. I imagine a team approach, with the Global EHS Director working with the senior company leadership, while the safety professionals on the battlefield work on employee and management challenges in facilities. I don't know if it will work, but this model and others like it should be toyed with and practiced in different ways; not because what we are doing now isn't any good, but because safety needs to continue evolving. Like Deming said, "It is not necessary to change. Survival is not mandatory" (brainyquote.com, n.d.b.).

---

*Deming*
W. Edwards Deming (yes, his middle name was pluralized; his parents somehow knew he was too big for a singular middle name) pretty much made modern industry what it is today through his philosophies. He did too much to go into detail here, but you should read a book on him. He has touched your life more than most people realize.

---

Survival? Really? Sounds a bit dramatic, doesn't it? The need to adapt professionally *is* dramatic, and needs to be taken seriously. Safety will continue to be about systems and engineering. As technology allows us to perform more mitigation on the high end of the hierarchy of controls, it phases out more reliance on the low end. It becomes easier and easier to justify taking an engineer who is talented at finding and reading standards, and placing them into the safety role. If safety professionals don't keep up, our profession will wither and become another hat that someone in the engineering department wears.

Okay, for those of you who are a member of the engineering department and you've been stuck wearing a safety hat, please know that I like you! We need you! But the collective safety profession also needs people who have degrees in psychology, chemistry, English (seriously), and philosophy (only half kidding). It is this diverse mix that allows safety to philosophically evolve as an art and science and keep up with the rest of the world.

---

*The Hierarchy of Controls*
There are a few different versions of the Hierarchy of Controls. I like a three-level model. At the top you have engineering and design; this means doing things like designing hazards away, or replacing a hazardous chemical with something less hazardous. Next, you've got administration; this includes training, job rotation, warning signs, etc. The lowest (and least desirable) level of control is personal protective equipment, or PPE.

---

The diversity of safety professionals extends past educational backgrounds. When I began my career a decade and a half ago, most of the conferences and educational seminars I attended were full of middle aged white guys. All any safety professional has to do nowadays to see how the age, sex, and ethnic background of the safety profession has expanded is look around the room during a conference. The development of social media, coupled with the expanding diversity of student bodies, allows a level of interaction that gives all of us the opportunity to broaden our horizons. A group of middle aged white guys tends to see the world in one similar fashion, just like a group of twenty-something women with brown skin tend to see the world in another similar fashion. The future of safety will grow most effectively when people of diverse backgrounds intermingle to share thoughts, ideas, experiences, and perceptions.

When I relate stories about my childhood, I oftentimes have to remind people that I grew up in a rural Minnesota town where ethnic diversity was measured by whether your genealogy was Polish, German, Norwegian, or Russian. And my hometown was a more diverse town than many of those around it. Local legend is that the railroads plotted the towns where stops would be, then populated those towns with people from different European countries with different religions so that the towns wouldn't intermingle too much. I don't know if that is true or if it was just that people who speak the same language and go to the same church tend to find each other, but many of the towns where I grew up are reflective of this non-diversity. When I left home and went to U.S. Army basic training and then college, I discovered a whole new world of thoughts, beliefs, language, and food. It is this giant, creative, humming jumble of human experience and diversity that will be the catalyst for change in the safety profession as we march further into the 21st century. Change is coming to safety, whether we seek it or not. The role of safety leaders is to guide the changes that occur in a manner that stewards the profession responsibly into the future.

# 3    Written in blood
## The death of General Patton

George S. Patton was an ass. He was unlikable and did some really awful things; after the war he wrote horridly anti-Semitic entries into his diary about the Jews in concentration camps (*New York Times*, 2015). A brilliant battlefield commander with incredible intelligence and energy, Patton was one of the most important American Generals of World War 2, but the guy was an ass. By dying the way he did, he was also a safety pioneer.

As a young man, Patton struggled with academics, but excelled at military drill and athletics. His first assignment in the army was with a cavalry unit in 1909 (when they still rode horses). He was an Olympic athlete (modern pentathlon) in 1912, placing fifth in his event. He would have competed in the 1916 games as well, had they not been cancelled due to the outbreak of World War 1. He was such a respected swordsman that the army adopted his sword design as standard issue in 1913, and he introduced new sword fighting styles for mounted soldiers (Axelrod, 2006).

---

*History and safety*
It is an unfortunate truth that safety laws are not passed until there has been bloodshed. If you're wondering why certain rules are in place, remember that even the ones that don't make much sense to you are the result of people getting hurt and/or killed.

---

It was in 1916 that Patton was assigned command of troops who were trying to find and eliminate Pancho Villa, a Mexican who had raided a town in New Mexico, killing several Americans. During the search for Pancho Villa, several Dodge touring cars were used as cavalry vehicles by the US Army, rather than relying on horses. The cars proved successful and Patton performed well; this combination of man and machine would change the world (Zaloga, 2010).

During World War 1, Patton established a tank school and became a pioneer in tank battle tactics. He rose in rank from Lieutenant to Colonel by the end of the war, and had gained a reputation for leading his troops from the front.

Patton was highly decorated, performed reconnaissance personally, rode into battle on top of a tank, and gained invaluable experience in tank fighting. After the war, he lobbied for a strong tank corps that would be its own fighting force rather than a tool to support infantry troops (D'Este, 1995).

Patton wasn't geared well for peacetime duty but he tried to make the most out of the years between wars, writing instructional manuals and developing battle tactics. In 1937, he predicted that the Japanese would launch a surprise attack on Hawaii. Unfortunately, he also devised a plan to intern Japanese-Americans in the event of such an attack (D'Este, 1995). He chomped at the bit for more wartime action, falling into bad sorts as peace dragged on. His frustrations manifested themselves in the form of drinking too much and having an affair with a niece by marriage who was much younger than him (there is some debate amongst historians whether Patton's claims of the sexual affair were real or contrived to boost his ego) (Axelrod, 2006).

When World War 2 broke out in 1939, I picture Patton jumping up and down, clapping with excitement. He was put in charge of tank units and was instrumental in building U.S. armored forces in preparation for war. He even learned to fly so he could view maneuvers from the air to better visualize how to choreograph large numbers of fighting vehicles. He was assigned to North Africa, where he performed brilliantly and was an inspiration to his forces (Axelrod, 2006).

After North Africa, he moved on to Italy, where he again commanded brilliant victories. His personality, however, became a problem. He killed an Italian civilian's mules who wouldn't get out of the way of an armored column, then assaulted the mules' owner with a stick. He did not punish soldiers under his command who committed a war crime by massacring seventy-three unarmed POWs. Most notably, he slapped and verbally berated two soldiers who had been moved to a hospital to recover from battle fatigue, or what we now call post-traumatic stress disorder (Axelrod, 2006).

Some historians postulate that Patton became unstable as a result of brain trauma, and his personal history certainly seems to indicate that there was mental instability. Because mental illness and brain injuries were not understood in the same way they are today, Patton likely never sought out medical advice in regard to his emotional upheavals. When Dwight Eisenhower found out about the slapping incidents, he removed Patton from combat command for nearly a year, using him as a decoy to make the Germans think he was preparing for an offensive landing and invasion, while Eisenhower secretly planned the D-Day landings (D'Este, 1995).

Patton did see more combat in Europe and when the war ended, he was again assigned to peacetime commands that he saw as unfulfilling. Here's where safety comes into play. Patton was by all reports depressed. The man was built for war, and war was now over. He felt that he had no more use and would be retired or dead by the time the next war broke out. He didn't know how right he was (Axelrod, 2006).

> *Seat belts*
> Seat belts were required to be installed in all new vehicles beginning January 1, 1968; the bizarre exception being buses (US Department of Transportation Highway Traffic Safety Administration, 1968). New York was the first state to require seat belt use, in 1984.

On December 8, 1945, Patton was on his way to a pheasant hunting trip in occupied Germany. Patton rode in the rear seat of a 1938 Cadillac Model 75. The car crashed into a military truck at low speed, and Patton was thrown forward into a partition between the front and back seats. The best information I can find (I have never seen Patton's car in person) is that seat belts were not installed in the vehicle. Available or not, seat belts were not worn. If Patton had been properly belted in, he would likely have not suffered any injury, let alone a broken neck that caused paralysis and led to death by a pulmonary edema twelve days later (Axelrod, 2006).

Patton was just one of many soldiers who died in preventable accidents after the war, and the military realized the importance of strong safety programs. The high number of peacetime casualties, including Patton's, led to a focus on safety. The military has pioneered many safety philosophies, from requiring seat belt use before there were any laws regarding the subject, to risk assessment methods and system safety tools, in an effort to reduce preventable casualties. Who knows what would have become of Patton after the war if he had lived. Perhaps the history of the Cold War would have been very different, or he would have devised different tactics to be used in the conflicts that followed World War 2. Or perhaps he would have become such an unsavory figure in retirement that there never would have been a movie made about him, turning him into a folk hero.

Like all safety standards and rules, military accident prevention is written in blood. Throughout the book, I will look back on other events that have motivated change to occur in safety. History is a wonderful teacher, and it is important for those of us who dedicate our professional lives to safety to understand the blood that was spilled for the standards we now rely on to be written.

One of the changes that safety professionals must continue to push for is to change this history of waiting for injuries and deaths to occur before making safety a priority. We have the knowledge and tools to anticipate most injury scenarios that are likely to occur when a new product, experimental manufacturing method, etc. is introduced. It is time for us to teach policy makers how to use that knowledge and those tools to make safety proactive across all facets of life.

# 4 Ergonomics
## Welcome to the jungle

A note on this chapter before I really dig into things… The safety profession, like many other vocations, is a practice that involves both art and science. We use art in our writing, designs, training packages, and many other aspects of the profession. We use science to find answers to problems, to ensure that artful design has function, and to examine facts regarding the human body, psychology, and performance factors; science makes our profession viable. To be the most effective safety professional possible, one has to subscribe to sound science and follow the principles found therein. That is why safety professionals are best served to accept the basic biological tenet that human beings are an animal of nature, evolved from genetic ancestors that were also animals. I am not asking you to discard religious belief if that is part of your life; only to accept the science as well.

Once we humans crawled out of the jungle and stood up on our feet with our oddly backward-turned hips and our oversized brains, filled with 100 trillion neural connections firing away to give us reasoning and logic skills, our natural environment instantly changed from a small patch of savannah on the African plains to the entire planet (Zimmer, 2011).

About two million years ago a baby was born that, as perceived by its parents and tribes, didn't look quite right. As the baby grew, it was obvious to the rest of the tribe that it was smarter, and could work out complicated logic problems. As the child entered adulthood, it began making better tools, and finding complex patterns in nature that made hunting and gathering more efficient and successful. The tribe was better off because of the existence of this individual, and the individual procreated, passing more of his or her genes along. This baby was the first Homo habilis (Schrenk, Kullmer, & Bromage, 2007).

---

*Cumulative trauma disorders and musculoskeletal injuries*
Ergonomic injuries are generally referred to as cumulative trauma disorders or musculoskeletal injuries. The terms, when generally used, are interchangeable and simply mean an injury that was caused partly or in whole by ergonomic issues.

---

It is likely that cumulative trauma disorders arose in conjunction with the rise of humanity. As our Homo habilis ancestors spent more time hunched over, chipping away at chunks of flint or bone to make spear heads, they felt tension in their shoulders just like we do now at our computers. They probably got a chip of flint in their eye now and then as well. It is easy to imagine that labor divisions inevitably began as those former hunters who had wrecked their eyesight with a chip of stone became dedicated weapon makers for the tribe. As hunts were more successful and rudimentary agriculture came into existence, skins needed to be dried and processed to make blankets, robes, and other useful items while food was harvested, stored, and prepared. People advanced farming until it became a science, and more divisions of labor occurred as some tribe members began making plows, while others dedicated their time to studying crops, growing new foods, and domesticating animals. Our species evolved along with these changes; it has taken untold years for civilization to reach our present day state from our first Homo habilis ancestor.

The problem with our present day state is that we aren't built for it. We evolved for a state of nature. There were no offices on that long-ago African savannah where our Homo habilis ancestors made their debut. We are engineered for working seven days a week, yes; but our "jobs" were much different before we started dividing labor.

Our minds have wandered further from our natural state than our bodies, so the best way to think about what our ancestors' natural lives must have been like is to think about your most basic physical needs and wants. Based on what my body tells me on a typical day, here is what I infer my ancient relative, Cave Scott, would have experienced…

Cave Scott would have woke up early in the summer but later in the winter, keying off the brightness of the morning sun. On a rainy day, he would have slept in, perhaps enjoying some extra "cuddle time" with Mrs. Cave Scott. Once awake for the day, Cave Scott and his lovely wife would have enjoyed some berries and nuts for breakfast, gathered over the course of previous days, while they watched Cavey Jr. play under a tree. The day would have been spent gathering stuff to eat, climbing trees, swimming, napping, eating, and cuddling more (men think about sex once or twice per hour; about as much as they think about napping or eating and *way* more than I think about work on a day off) (Fisher, 2011). Because we have a relatively sparse amount of fur and the ability to sweat, Cave Scott would also spend a significant amount of time on foot chasing down animals to eat (Jessen, 2000). Researchers believe that our endurance abilities allowed ancient humans to chase after prey until the prey animals succumbed to heat exhaustion.

Many of these activities that the natural human did are at odds with our current lifestyle. Because our brains have developed faster than our bodies have been able to adapt, we nowadays force ourselves into a very unnatural environment. George Costanza on *Seinfeld* could have argued that he was just trying to get back to a more natural state when he set up a nap area under his desk. This concept, that our bodies are still better engineered for a state of nature, is key to finding solutions to ergonomic riddles.

Forget a person's seated posture. Forget about sitting on an inflatable ball. There are no chairs or yoga balls in nature. Unless someone has a medical necessity for a chair, employees should stand (LifeSpan, n.d.). I'm standing while I write this. The reason more and more research is showing that sitting is terrible for our bodies is because sitting is not something we do in nature. Humans are meant to move; we are designed specifically for it. Homo sapiens and Equus feres caballus (horses) are two of the very small number of species that produce large amounts of sweat (Jessen, 2000). Why is this, and what can it tell us about ergonomics?

Horses need to move to be happy and healthy. They need to move *a lot* (Harris, 1994). That amount of movement causes the body to heat up. Sweat cools it back down. Humans also need to move to be happy and healthy. We need to move *a lot*. In modern times and with a regular job we can't move as much as we normally would in nature, because then we'd also spend a large portion of the day napping, and would do more work in the dawn and dusk hours that are more cool and comfortable for us (Bramble & Lieberman, 2004).

What would a zoo exhibit look like with a natural human? Forget about that goofball who goes around to zoos and lives in a glass apartment. I'm talking about a true natural state. There would be a stream for fresh water and fish, some trees and grass, and tons of room to roam around (Hopkin, 2005). If you were outside all day, think you'd just stay in one spot? We humans would probably need the largest enclosure of any animal in the zoo to stay happy.

---

*Static versus dynamic*
A static posture or state is one in which the subject moves very little. A dynamic posture or state is one in which the subject moves around frequently.

---

This need to move that we share with horses plays a huge role in modern ergonomics. Workplaces tend to force humans to remain static for a large portion of the day. In nature, when our bodies are static, we start to get tired. Think of it like this, how often do people randomly fall asleep when they are sitting still? When a human sits still, he wants to get something to eat and occupy his mind, then take a nap (Ehrman, 2006). Imagine trying to force horses into a work situation like we do to humans. They'd go nuts and they would physically whither. We do the same.

Thinking about humans in our natural state tells us everything we need to know about ergonomics. The first key is to be dynamic; just move around. For office workers, this means standing workstations and getting out of the cube to walk frequently. Because we evolved to sweat, we require lots of water to drink (Jessen, 2000). Encourage office workers to drink lots of water, and they will pee more. Employees may not want to take time away from work to just walk around and move their body, but they will definitely walk to the bathroom when they have to pee.

Get rid of printers in offices or cubes, and force people to walk to a printer. For that matter, prohibit anything else that keeps people anchored. Prohibit coffee makers in individuals' offices; people can walk to the shared coffee station. Get rid of space heaters. If people are cold, they should take a walk and move their body to warm up.

What about employees on the production floor? Get them moving, too! The best way to prevent ergonomic problems in production areas is to change up the jobs people do on a regular basis. We aren't made to repeat the same actions over and over for an eight or twelve hour shift, five days a week (OSHA, 1993). It is no wonder musculoskeletal injuries are so pervasive!

---

*OSHA*

The Occupational Safety and Health Administration (OSHA) is the federal agency in charge of making workplace safety laws. We'll talk about the history of OSHA in another chapter. States can have their own OSHA agency, in which case, the state laws have to protect workers as much or more than federal OSHA standards.

---

There are a few jobs where dynamic movement throughout the day simply cannot be made practical given our current technology. Truck drivers, for instance, cannot feasibly pull over every hour or so and run a couple of laps around their truck. For these jobs, we have to rely on our current paradigms of thought about ergonomics, but for the vast majority of the other jobs out there, a shift in thinking about the physical needs of the human animal will pay large dividends.

As technology advances, safety professionals will have more opportunities to return humans to something resembling their natural state. Self-driving vehicles are on the horizon, which means truck driver ergonomic calculus will fade and disappear in time (along with truck driving as a career). More of the highly repetitive tasks in workplaces will be performed by robots, leaving jobs to humans that require lots of dynamic movement that is dependent on changing environmental variables. In other words, what the jobs robots won't do are those jobs that are always changing, and require an adaptive mind to change the method and type of physical work performed. It's going to be a while before a computer can do the work of 100 trillion neural connections. If technology means that labor shifts from being a producer of goods to being a monitor of production devices, safety professionals can keep employees ergonomically healthy by encouraging dynamic movement during the monitoring tasks. Employees may not always like the answer to the problem, but the answer really is to get on your feet and move.

The CDC reports that more than one third of adults in the United States are obese (Centers for Disease Control, n.d.). This is a parallel to ergonomics, because the obesity rate is affected by a rise in sedentary lifestyles. Since most safety professionals have the word "health" somewhere in their title or degree, the obesity epidemic becomes part of our job. While we cannot control the diet of the workforce to a great degree, we can remove factors that give rise to a sedentary lifestyle. Our discussion returns to a standing workstation. Standing burns calories (Livestrong, 2015).

There are definitely drawbacks to standing. People get tired, it can be rough on the joints, and some researchers think it contributes to varicose veins (Mullis, 2012). The ultimate goal, though is *dynamic movement* rather than any static posture. Standing in one place all day is not dynamic, it is static. So any time we place an employee in a standing position for work, we need to also provide them with time and motivation for dynamic movement. Watch someone at a standing workstation, and you will see that they shift their weight and make other movements frequently; it is easy to get them to move even more with some gentle prodding.

Safety nerds can also decrease obesity rates among workers with the limited control we have over their diet in the workplace. Reduce or eliminate unhealthy snacks in vending machines. If you can't get rid of pop machines and candy bars, at least offer and encourage healthy alternatives. Meet with the cafeteria vendor or manager if you have one, and begin talking with them about eliminating or reducing unhealthy foods. I'm not saying that you have to be the pizza police, but most cafeteria vendors or kitchen managers can find the healthiest options for their recipes or make sure healthy alternatives are offered. Many employers provide budgetary resources for a health committee; use these resources to educate employees about nutrition. It is surprising how many people are nutrition illiterate.

Think back to Cave Scott. He was eating fresh foods with little to no processing. Cave Scott might have cut up an antelope quarter, rubbed spices on it, and cooked it over a fire; eaten some berries and nuts that were baked over that same fire with a wheat crust to make a crude desert; or have dried or smoked fish to preserve it for a later meal. What he wouldn't have done is synthesized molecules, refined sugars, or boiled the syrup out of corn. I'm not some kind of food hippie, believe me. I have a brother who lobbies his local city hall to make sugary drinks illegal, and I think he's probably closer to nuts than normal on the sanity spectrum. I like a pop now and then because it makes my hard liquor taste better. I can sit on my couch while watching the Vikings and motor through an entire frozen pizza, then down some chips and salsa while I watch the Packers lose the night game, all while working on a few beers. But in general, I make modest daily efforts to maintain a reasonably healthy diet and get some exercise.

The more people move their bodies and the healthier they eat, the less likely they are to suffer from a musculoskeletal injury (Salvendy, 2012). Between improved technology, a more naturalistic, scientifically sound view of the human animal, and more of a focus on overall health, safety professionals have the opportunity in the 21st century to reduce ergonomic injuries to levels we can hardly imagine today. We can put the "H" back into EHS, and the dividends to the company will be significant. Reduced absenteeism, fewer medical insurance claims, and less turnover due to employees who die of health related conditions are all predictable advantages of focusing on the human as a natural animal. Welcome to the jungle.

# 5  Ptolemy, Freud, and Heinrich walked into a bar

But Einstein ducked.

Astronomy today recognizes, through observation and calculation, certain physical truths. For instance, it is accepted as fact that the sun is the center of our solar system. One of the early figures in astronomy who advanced the science significantly was a man named Claudius Ptolemy, a Greek citizen of the Roman Empire around the first century CE. Ptolemy used mathematics and the scientific method to write texts on the physical world, including the universe (Goldstein, 1997).

Ptolemy's writings and tables also include information on music, geography, and optics (Partch, 1979; Smith, 1996). Astronomers today generally agree that without Ptolemy, it is likely that other subsequent astronomical advances would have been delayed by years, decades, or centuries (Greene, n.d.). His significance to the field can't be overstated.

It is also important to note that Ptolemy thought the earth was the center of the universe. This important scientific figure, a man who advanced human knowledge in so many ways, was wrong about a fundamental piece of data that is now accepted as undeniable fact (Goldstein, 1997).

Sigmund Freud is a little more modern. He did his work with neurology and psychoanalytics from the late 1800s to the early 1900s. I was a psychology major, and a number of my classes began with discussions of Freud's theories and discoveries. When Freud was working with patients who had psychological problems, he basically looked for what a safety nerd would call the root cause. Freud realized that much of an adult's neuroses could be traced to events that occurred much earlier in life. Because of his work, we now are much more adept at treating psychological issues.

Freud advanced knowledge of cerebral palsy, brain anatomy, hypnosis, and our understanding of the unconscious mind. Without Freud, who knows where our understanding of mental illness and the mind would be today (Wollheim, 1981).

Sigmund Freud also believed that cocaine was a cure for many mental problems, and recommended it as a cure to morphine addiction. He attributed many neuroses to a child's sexual desire for their parent of the opposite sex (no, no, and more no), and thought all dreams were latent secret wishes (so apparently I wish that a silverback gorilla dressed as a clown will eat my legs while I am ice fishing. Sure.) (Wollheim, 1981).

Herbert Heinrich became a safety pioneer in the 1930s. His life and working period overlapped Freud's. While Freud was escaping Nazi Germany and quipping "What progress we are making. In the Middle Ages they would have burned me. Now, they are content with burning my books," Heinrich was performing research that has shaped the safety profession for decades (BrainyQuote, n.d.a.). It is time for safety professionals to accept once and for all that the way we practice safety today has to move beyond Heinrich.

Just like Ptolemy is familiar to astronomers and Freud is familiar to psychologists, Heinrich will need to be familiar to safety professionals for many years to come. But as Fred Manuele astutely points out in his book *Heinrich Revisited: Truisms or Myths*, much of Heinrich's ideas need to be recognized as inadequate to our current understanding of the art and science of safety (Manuele, 2002).

> **BBS**
> Simplified, behavior-based safety is the idea that you can modify employees' behaviors so they take fewer risks at work (Cambridge Center for Behavior Studies, n.d.).

Heinrich's influence is not always completely obvious to the safety profession. It has seeped into the "common knowledge" of safety insidiously; it rears its head every time a supervisor tells you that an employee should have known better than to stick their hand in there. Behavior-based safety exists because of, and is rooted in, Heinrich's ideas about accident causation (Heinrich, 1931). In BBS courses, people are still taught Heinrich's pyramid as if it is relevant (see the boxes for more information on BBS and the hierarchy of controls). Do you really mean to tell a vice president who is the manager of a very low-risk corporation that deals solely in office work that for every 300 minor injuries one employee will die? That's one hell of a paper cut. Don't even get me started on the hazards of paper clips. And staplers? People are now running in fear of them (let's just hope they aren't holding scissors).

Believers in BBS will argue that the accident pyramid is only a principle, and should not be taken literally. I argue that it does not even work in principle. In the seven years prior to exploding, killing eleven workers, and spilling nearly five million barrels of oil into the Gulf of Mexico, the *Deepwater Horizon* had an excellent safety record (Casselman, 2010; United States Coast Guard, 2011). There was no base to the pyramid; there were no stockpiles of minor incidents, and had there been, they likely wouldn't have ever been indicative of the problem that led up to the disaster.

Next time you are wondering about how much emphasis to place on Heinrich's ideas, sit down with a recently injured employee and tell them that there is an 88–95 percent chance the injury was their own damn fault (Heinrich, 1931). In any incident investigation, as the chain of events are linked together, you are going to find multiple instances where behavior was a contributing factor. Note, however, that Heinrich himself recognized the way to break those links and create a safer system was to change the environment to accommodate, work around, or negate human behaviors that contribute to incidents. While some of Heinrich's work has inspired safety philosophies that are detrimental to our profession, other aspects of his work have gone on to inspire safety philosophies that are based on sound engineering principles. This is why he is in the realm of Ptolemy and Freud instead of Rasputin (Heinrich, 1931).

---

*Human factors*

There are a ton of books and research papers on human error. Most of us think human error is simply mistakes made by people; but the real trick is finding the reason why someone made a mistake. We all make mistakes, and I believe that an entire realm of safety in the 21$^{st}$ century will be dedicated solely to examining human factors of design to reduce human error by improving the design of items.

---

We've got an opportunity here to examine human error. Human beings are incredible machines; do you know of any other single machine that can be useful when mowing the lawn, flying a space craft, and cooking an egg, all while making independent decisions and complex calculations? Robots aren't close yet, and not even the most complex computer has reasoning skills to do all these jobs while dreaming about new places to explore or inventing a new machine from nothing. Why, then, do we make errors; and does a human error mean the human was the root cause of the incident? More importantly, at what level is a single human error magnified to the greatest extent?

Humans make errors all the time (see my chapter on DAF). With a good safety design, we can take the lessons learned from human errors and apply them to improved design standards. The jerks of the world get a bug up their bottom because we do things like look at how many kids have been injured by falling off of a swing set onto the pavement, then change playgrounds to have soft surfaces; or we identify that a high number of people who are killed or gravely injured in motorcycle crashes weren't wearing helmets, and we push for helmet laws. It is often human error that causes us to fall off a swing or dump a bike on a corner with loose gravel; but it isn't always the human error of the injured party. Maybe Heinrich should have dug deeper to determine how many human errors occurred within a *management system* rather than focusing on individual employee errors, because in a management system, human error is magnified to a much greater degree.

Example one is the space shuttle *Challenger*. According to Heinrich, there is an 88–95 percent chance the crew's behavior caused the destruction of the orbiter. The investigation did indeed find that human error was the root cause of the explosion that occurred 73 seconds into the flight, killing the seven astronauts on board. A failed O-ring seal in the right solid booster rocket was the ultimate physical cause of the explosion, but a human error occurred within the management system. Engineers detected O-ring problems as early as the second shuttle flight in November 1981, over four years before the *Challenger* explosion, but management decisions allowed the potentially catastrophic condition to continue. Humans made the erroneous decision to launch in weather colder than the O-rings had been tested for, and the result was tragic. You might be saying to yourself, "That sounds like it fits into the 88–95 percent category." The problem with that assertion is that this was a human error within the *management system*. Failures of management systems must be considered separately from unsafe acts of employees because management decisions are typically reviewed to predict where they may contain errors. Management system failures are human decision errors on the part of those in control of employees, systems, etc. (Rogers Commission, 1986).

Example two is the case of every automobile manufacturer ever. When Richard Grimshaw was thirteen he was badly burned in a car wreck. Richard was the passenger in a Ford Pinto that stalled out on the freeway, was rear-ended, and burst into flames. The Ford Motor Company knew that there were design flaws that could lead to this type of fuel tank failure in a crash, but determined that it was less expensive to accept the design as it was and risk a predicted number of settlements. I know this reasoning sounds cold-blooded, but it is important to remember that the senior leadership team (SLT) of Ford at that time thought they were making the decision that would result in the greatest good for their shareholders (*Grimshaw v. Ford Motor Company*, 1981). I'm not going to beat up on Ford here, because every other car company has made similar decisions. There are tons of examples, whether it is the unfortunate Mr. Grimshaw, the people who died as a result of faulty ignition switches in GM vehicles (Valdes-Dapena & Yellin, n.d.), or those persons killed by runaway Toyota accelerators (Vlasic & Apuzzo, 2014). There was, in nearly every case of a vehicle recall, a management system failure that stemmed from *decision making* authorities within the company who should have known better.

In the above examples, a management system with a more robust focus on risk mitigation would have possibly eliminated the safety issue before it ever affected the operator. Safety professionals can take these examples, the extremes, and apply them to most other safety situations. The damage done to the organization in the above examples was grievous, even when taken separately from the personal tragedies of the individuals involved. As of this writing, NASA is hitching rides with the Russians when we need to get an astronaut in space. Can you imagine if Spain decided to send a few explorers to the New World on English boats fifty years after Columbus mistakenly stumbled upon America? That's what is happening now. For a car company, a recall is a huge drain on resources and reputation. I wasn't even born when Richard Grimshaw suffered his injury, but people my age still use the Ford Pinto as an example of poor decision making.

Today's safety professional has an opportunity that didn't exist in quite the same way even a few years ago. Because information travels so far so fast and has acquired great value, the importance of reputation management allows the safety professional to present the concrete facts regarding a hazard to managers, along with statements regarding the possible reactions by employees, community members, etc. to the mitigation activity (or inactivity) that management decides is appropriate. The curtain has been pulled back, and we can see who the Wizard really is. The job of the safety professional must shift; not to nail the curtain down, but to help make sure there isn't a humbug behind it.

Employees are going to make errors, sometimes really stupid errors. The result of a predictable human error should not be a disabling injury. Refer to the chapter titled "DAF" and recall that we have the duty to predict that if a person *can* make the error, at some point a person *will* make the error (no matter how stupid it might seem). People have bad days, people get mentally preoccupied, and not everything you, design engineers, or I, think is common sense really is "common sense." A great example is a fatality investigation I led. The deceased employee had a three-week old daughter at home. Being a parent myself, I know how much sleep deprivation becomes a factor in everyday life with a new baby at home. No parent should be blamed for their own death because their baby was upset and gassy the previous night.

# 6 The time I thought I killed myself with a sign post and a nitrogen cryogenic cylinder

I hate to share stories like this, but I promised fun in the introduction. Seeing your life flash before your eyes is fun, right? Well, maybe fun for you guys to read about, anyway. I knew I should have promised nothing but serious discussions and zero stories about times I was an idiot.

The facility was changing addresses. We built a new building about a mile down the road from the old place, and were in the middle of moving in. One of the tasks I had was to determine where our evacuation meeting points should be, then get them set up properly. The facility manager and myself had picked two perfect spots for evacuation meeting points, and decided that we would mark each with a big blue sign. The first meeting point sign would have a big "1" on it, and the second would have a big "2" on it.

I ordered the posts, the signs, the stick-on numbers, and made sure we had the tools to get everything installed. When the materials arrived, the landscapers were still working on the grounds around the building, and it was going to be a week or two before the signs could be posted. I had limited storage areas for the materials, so I stuck the smaller items in the desk drawer I was given for temporary inventory storage. The posts, however, were eight feet long and wouldn't fit in my small storage space.

---

*Cryogenic cylinders and relief valves*
Every cryogenic cylinder needs to be able to relieve pressure. Using the combined gas laws (I heart physics), we can see that if the temperature of the material rises to create a gas, the gas will continue to expand in volume if the temperature continues to rise. This causes the gas to escape through the pressure relief valve. Happens all the time, and people familiar with cryogenic storage systems are used to it.

---

I conferred with the maintenance crew, and decided to store the posts on a large shelf in the chemical storage room. We only had a few chemicals in the inventory; full orders were on hold until we were completely moved in, so there was plenty of empty space alongside the two flammable cabinets and the five hundred pound nitrogen cylinder.

The air conditioning was not yet fully operational at this point, and the warm air caused the cryogenic fluid in the nitrogen tank to regularly convert some of its volume into gas and build up pressure until the cylinder's relief valve opened for a second or two. Had somebody related this vital fact to me, I'd have no good story to tell you right now.

So there I was in the chemical storage room. Roger, a maintenance technician, was in the far end of the fairly large room working on something with a contractor. I had the two eight foot posts in my hands, and was trying to maneuver through the room to get them onto the storage shelf when I gently tapped (no exaggeration, it really was gentle) the nitrogen cylinder. The vibration was just enough energy added to the gas pressure to open the pressure relief valve for one second.

As I had no knowledge that the valve was opening for a second at a time every three or four hours, I thought I had hit the stem of the primary valve and caused a release. In the one second the valve was open and there was a loud hissing of gas, I had the following thoughts in this order:

1   Holy buckets, I'm going to die today.
2   I could drop these posts and run, but I need to keep a clear path for Roger and the contractor to get out.
3   I wonder how long it will take the cylinder full of nitrogen to displace the oxygen in the room, and will we suffocate before the cylinder is empty?
4   This is going to get cold.
5   I'm glad I have life insurance so my family will be financially OK.
6   Why did Roger just start giggling?

At that point, the hissing stopped. Roger broke into a full laugh, having seen the look on my face when I thought the end was nigh.

"It's a little warm in here today, so it just does that," he said, wiping tears of laughter from his eyes. I thought the contractor was going to pee his pants, he was laughing so hard.

---

*Nitrogen*
Nitrogen makes up over ¾ of the atmosphere; we breathe a lot of it (Emsley, 2011). Large releases kill people by displacing oxygen and causing suffocation.

---

I don't remember my exact reply, but it was something to effect of "I thought I was a goner," with a lot of swearing mixed in. I gently placed the posts on the shelf, had a nervous laugh about the situation with Roger, and then went to find out how long it would be until the HVAC was cooling the building enough to stop the cylinder from building up pressure and relieving. I had a nice scotch that night with dinner.

> *HVAC*
> Most safety professionals have some familiarity with heating, ventilation, and air conditioning basics. We typically use this basic knowledge when evaluating or designing ventilation methods that will reduce chemical exposures.

My primary take-away was that in the emergency, I didn't panic. I thought about throwing down the posts and bolting, but I didn't. My brain was analyzing things, looking for a sensible solution while collecting data related to the situation. I gained an appreciation for why people sometimes seem to freeze up in moments of extreme duress; they are oftentimes analyzing their predicament. I have used that experience to help guide me through investigations and incident analyses.

I still hate it when Roger tells the story to new people at work, though.

# 7 Common sense skills / common sense kills

I grew up in rural Minnesota. I can tell you ten things about a walleye without having to think. I know where to shoot a deer so it drops in its tracks; and if it doesn't drop in its tracks I can track it through the woods for miles. I know how to drive a sub-compact car with bald tires through a blizzard, and I know when it is a really bad idea to swim in a lake for any one of a dozen reasons. These things are all common sense to me.

I have a cousin who grew up in the Ozark Mountains of Southern Missouri. Yep, he's a hillbilly. He knows how to stick his hand in a hole in the bottom of a river and pull out a catfish the size of my neighbor's German Shepherd. He knows which bullfrogs have legs that make for good eats, and how to catch the buggers. He can drive an ancient Jeep up what looks to me like a 90 degree slope without the beer cooler falling out the back, and he knows when to avoid swimming in a creek because there might be snakes. These things are common sense to him.

Common sense is shaped by an individual's geography, language and dialect, age, social group, economic status, and a myriad of other factors. What is common sense to me is not common sense to my cousin from the back hills. I grew up in a time, place, cohort, and using a dialect very different from most other people on the planet (uff-da!). My common sense has been shaped by influences unique to me. I took this to heart over twenty years ago in basic training when I was thrown into a platoon with about thirty other people from across the USA. Every one of those guys (only men were allowed to be combat engineers back then) thought about things in a completely different way than I ever had and they used different words than I ever had. We all had to learn each other's common sense, which meant none of our sense was common.

---

*Police car safety*
A couple of items of interest I have learned over the years. A lot of cops who write other people tickets for distracted driving use their laptop computer while they drive, and there are special car seats designed just to try to reduce ergonomic injuries to cops on patrol.

Take two individuals from different backgrounds and put them into this situation: A police officer pulls them over late at night and asks them to step out of the car. Each of them has something in the car they'd rather not have the officer find. The police officer asks each person, "Where are you going tonight?"

The first person remembers the police officer that stopped by his elementary school to talk about bike safety. Then he thinks of the officer who helped him find his dog, and the other officer who made sure he was OK after a fender bender on the freeway. He answers, "Headed home after poker night."

The second person remembers when an officer came to his house and arrested his older brother who had been home studying all night. Then he thinks about the officer who unreasonably stopped him and questioned him for twenty minutes one night when he was walking home after watching a high school football game, and the officer who called him a rather nasty name for no good reason. He answers, "Jail, once you search my car." Both of these people responded according to the "common sense" they had acquired.

When I worked for a grain handling company, a huge pile of interns were hired each summer. These bright eyed college kids were anxious to get their hands dirty and wanted to learn as much about agriculture as they could. Because agriculture is becoming more of a big business, many of these young people did not have farming backgrounds and sought a career in agriculture simply because of the opportunities it offers. I received an injury report that an intern cut his forehead and needed stitches after a post pounder slipped off a post and whacked him in the head.

---

*A brief chronological outline of my safety career*

- Safety Officer for a prison
- Consultant
- Safety Specialist for a shoe manufacturer
- Corporate Safety Manager for a grocery store chain
- Loss Control Consultant for an insurance company
- Safety Specialist for a grain company
- Safety Engineer for a high-tech company

---

Before I go any deeper into the story, I should tell you that you are part of a common sense experiment. Some of you, given the limited information above, can picture exactly what happened. Some of you might have a vague idea, and some of you don't have any clue what a post pounder is. The manager who reported the injury wrote the cause as a "lack of common sense." Do you have the common sense it takes to pound a post into the ground with no instruction?

A post pounder is a tubular piece of heavy steel with an open end and a closed end. The closed end is thick with metal so that it has significant weight. The user slips the open end of the tube over the top of a post that will be pounded into the ground, then lifts the tube and slams it down into the post over and over again to sink the post into the earth. The intern was injured when he lifted the pounder too high, it slipped off the post, and the weighted end smacked him in the head.

Accident causes like "lack of common sense" or "should have been more careful" bug the living crap out of me, so I called the manager on the phone.

"Hey, Troy. Tell me more about this kid who split his head open."

"Well," Troy said, "he's one of our interns. He's usually pretty sharp, but he really had his head up his ass this morning."

Thinking to myself that if his head was up his ass, it would have been protected and the pounder wouldn't have been able to make contact, I asked "How much instruction did he receive for this job?"

"I don't know. I sent him out there with Jim. I think Jim showed him where the posts needed to go and dropped him off with everything he needed."

"How much experience does this kid have with jobs like this? Has he pounded posts in the ground before?"

"Probably not. I guess I should have given him a helmet or something."

I spent the next ten minutes lecturing Troy on why all employees need to receive instruction for a new task. I also told him that I was changing the incident cause to a "management deficiency." Troy went nuts, and talked about how using a post pounder is common sense. I told him that it seemed like common sense to me that a manager should know that a new employee needs to be shown how to do his job safely.

That same summer, another intern rolled a pickup truck on a gravel road. Thankfully, she was not injured, but she did receive a good scare. The truck bed was full of heavy jugs of herbicide that were ejected in the crash, but luckily did not break open and cause a spill. The manager at that location listed the cause as "did not drive safely."

Our discussion was very similar to the one I had with Troy. The manager argued that everyone knows that when you've got a heavy load in the back of a truck and you're driving on gravel, you go slowly and are extra careful on turns. I asked what kind of car the intern drives, and the manager described something small, two-door, and front wheel drive. When I asked if the intern had *ever* driven a pickup truck before or if she had experience driving on gravel roads, I got a heavy sigh for an answer. He didn't know. I told the manager that it wasn't the intern's common sense that was lacking, it was *the manager*'s common sense that was lacking. I changed the incident cause to "management deficiency." Just a side note that the load technically should have only been driven by someone with a commercial driver's license. The manager and some of the upper leadership of the division had a separate discussion about the common sense involved with that aspect of the incident.

In England, there is a cartoon named "Peppa Pig." Peppa Pig is about a cute pink pig, her family, and her animal friends, and the show is exported to other countries like the USA and Australia. Some of you might have watched it with your kids. Peppa is a sweet little girl of four years old who enjoys some typical pig activities and some typical human activities (as any good cartoon piggy should). Peppa has had some safety controversies through the years, and they involve this elusive cat we call "common sense."

---

*Bike helmets*

When I was in sixth grade, I crashed my bike so hard I spent days in the hospital with a bad concussion. There are about two days of my life I don't remember. I wasn't wearing a helmet; nobody did back then (it was 1987, the year the Twins won their first World Series). Bike helmets save lives. Between 1975 and 2013, bicycle deaths dropped 26 percent. (*Helmet Related Statistics from Many Sources,* n.d.).

---

In the first two series of the show (the British quaintly call them "series" instead of seasons), Peppa's family did not wear seat belts in the car; characters on bicycles did not wear helmets. After complaints were lodged, the scenes were re-animated to include seat belts and helmets, and future seasons included use of these safety items. If you are one of those people who see safety nerds as getting involved in everyone's business, you're going nuts right now. In return, I'm laughing at you and telling someone about why you should have worn a helmet at birth, because your mom was standing up when you were born. Ha ha!

Peppa Pig was part of a much more serious controversy involving common sense in 2012. Common sense in the UK is that spiders are not dangerous. Of the over 650 species of spider in the UK, there are only a dozen or so that can hurt a person with their venom. None of them are normally deadly, although I did find a reference to a woman who died after being bitten by a spider called the "false widow." Because spiders in the UK are pretty much harmless, the Peppa Pig episode *Spider's Web* centered on informing toddles that they do not have to be afraid of spiders. Peppa plays with a spider named "Mr. Skinnylegs," picking him up by a thread of his web. It's a cute episode.

Remember that I mentioned Peppa Pig also airs in Australia? Australia has some of the most deadly spiders in the world. Children with common sense are afraid of them. In the UK, Mr. Skinnylegs is cute and helps keep indoor insect populations low. In Australia, Mr. Skinnylegs can send you to the hospital for anti-venom or even kill you. In 2014 in Sydney alone, according to *The Telegraph*, paramedics responded to 319 spider-related incidents. The episode featuring Mr. Skinnylegs aired once in Australia to a collective gasp of horror by parents with common sense around the country. It has never aired there again.

I'm a cat person. Any animal that knows how to use its own bathroom is alright by me. I don't want a codependent dog that idolizes me and has to go outside for a comfort break when it's twenty below zero out then wants to crawl into bed with me. It's too much pressure. I like a cat's attitude of independence coupled with good doses of cuddling and play. And, I hate the idea of walking around after an animal scooping its poop off the sidewalk or my neighbor's lawn. I changed my son's diapers because I knew someday he'd be potty trained (like a cat), but the idea of strolling down the sidewalk with a plastic bag of feces tied to my wrist is just foul. Back to cats… Most people, even those who don't like cats, know that when a cat purrs, that means the cat is happy. Common sense, right?

Wrong. Purring does indicate happiness in a cat *most* of the time. But cats also purr when they are afraid or in distress. Cat behaviorists think this is Kitty's way of calming herself when she's under stress (Lyons, 2006). I knew of a cat that roamed the courtyard of an apartment building I lived in during grad school. She had long, thick hair and was always happy to approach the residents as they walked in and out of the building. The first time I met her I was moving in.

I approached her and held out my hand, but one of my new neighbor's said, "Don't let her fool you. She suckers you in to pet her, then she attacks your hand."

"Really?" I asked. The cat seemed so nice.

"Yeah, she pretends to like it. Just about the time she really gets into it and purrs, she scratches you and bites. I hate that evil cat!"

I tested the story out. Sure enough, the cat let me pet her, and a few seconds after she started purring, she turned into Mistress Chompandclaw. It hurt. I did some research, and found out that a lot of cats have sensitive skin. They want the social interaction of petting, but after a short time it begins to hurt them. When this cat purred, she was beginning to feel pain and her purring was a reflexive action. The biting and scratching was her way of telling me she couldn't take anymore contact. After that, I would pet her for one or two strokes, then continue on my way. My neighbor's common sense told him that when a cat purrs, she is happy. His common sense also told him that when an animal pretends to be happy but then attacks you, the animal must be evil.

Now think about your favorite elderly relative from your childhood. Might be a grandparent or a great uncle. One of my favorites was my great uncle Albert. Remember the cousin I mentioned earlier in the chapter? Albert was his grandpa. Albert's world view was beyond conservative. He would have made Sarah Palin look like a bleeding heart liberal. Even though I was a young kid during the years our lives overlapped, I knew that he was a bit of a radical. One of the many things Albert liked to do was talk about people who should be shipped out of the country, always based on simple "common sense" observations.

Albert had a really cool wood shop in the shed behind his house. It leaned heavily to one side and there were birds who nested in the roof. They'd come flying in and out of holes in the rotting soffit, and people always knew when they had eggs or chicks, because they would dive bomb anyone walking into or out of the shed. Albert was a talented wood worker and made all sorts of projects in his shop, everything from rocking chairs to toys for his grandkids.

One day, Albert cut off about half of his right ring finger while using his band saw. The doctors were unable to reattach it, but Albert adjusted well. He was back to work in his shop in no time. One night while I was visiting, listening to Albert tell me about how I needed to shed my Yankee ways, he started talking again about people who should be shipped out of the country. My grandpa, who was Albert's younger brother, chimed in with "They should start with people who use up all the Medicare money because they don't have enough common sense not to hurt themselves."

I heard Albert say once that common sense wasn't common, and he was right. Just not for the reasons he thought. Common sense is not just uncommon, it simply doesn't exist. It is unique to every person because of their frame of reference. As safety professionals, it is our job to present this message to managers and employees, because a reliance on common sense hurts and kills people. As technology advances, there may be a greater separation in each generation's frame of reference as young people grow up with more screen time and less time experiencing the actual world. The generation that is watching Peppa Pig right now will have a completely different frame of reference on which to base their view of the world than you or I had. They will think of something as common sense that you or I believe is novel.

The term "common sense" needs to be retired from safety. It doesn't exist. It never has and it never will.

# 8 Everything I need to know about behavior-based safety I learned from my cats

Losing an argument to Scott Geller was the highlight of my day. It was the Minnesota Safety Council annual conference, and I found myself sitting at a lunch table with Scott. It was pretty exciting to have an opportunity to meet Mr. Geller in person, and the Safety Council served a tasty lunch. I couldn't wait to hear what a guy like Scott would talk about over roasted chicken. I always felt a kinship with him, another Scott G. in the safety world.

---

*E. Scott Geller*
E. Scott Geller is a superstar in safety. He's written a ton of books, and really has advanced the science of safety quite a bit through his research and innovations. While I disagree with Mr. Geller about how much emphasis should be placed on behavior-based safety, I have the utmost respect and admiration for him. I highly recommend reading his work.

---

Scott asked me what part of the job I liked the most, and I told him it was probably training employees. He corrected me, saying that employees should be educated. I nodded, and agreed that in most cases, educating was better than training. "Now and then," I added, "I'd rather just train them." Scott countered that education was *always* better.

I have vast respect for Scott Geller, and I think that many of his teachings are of huge value to safety professionals. But I also like to argue more than is healthy. I thought about fall protection. For those readers not in the safety world, a PFAS is a personal fall arrest system. "If I need an employee to put on a PFAS," I said, "I want to train them. I don't care if they know that when they fall they'll accelerate at nine-point-something meters per second. I need them to recognize they are four feet or more above the next level, and then step-by-step inspect and put on their system. It's more training than education." Scott Geller pointed out that I was educating them to understand when the system was needed, how to care for the system, and what could go wrong with a damaged harness.

I nodded, giving round one to the intellectually superior Scott G. "Okay," I continued, "what about fire alarms or tornado warnings?"

Scott asked, "What about them?"

"I want to train—purely train like one of Pavlov's dogs—employees to go to shelter when they hear a tornado siren, or to exit the building when they hear a fire alarm." Scott mulled this over for a second, and I continued, "I have a cat named Lucky. I've trained him so that when he hears a tornado siren, he runs downstairs to the basement. I want employees to do the same thing. No thought, no analysis, just stimulus and response."

I like to think I shined for a moment while Scott conceded that in cases like alarms, there may be some room for training. Then he asked how I had trained Lucky. "I used his natural behavior, and built off it," I said. "Lucky never has liked the tornado siren, and there's one right outside our house. When I decided to train him, I worked on rewarding him for doing what came naturally—hiding from the siren. When the siren sounded I picked him up and took him downstairs. Once we were in the basement and the siren was barely audible, I gave him a treat and pet him until he was relaxed and purring. I continued doing this until he was meowing at me to come with him and trotting to the stairway when the siren sounded. I wasn't convinced that he would go there when I was gone at work, so I left work early one day when there was going to be a siren test and spied on him." In the town where I lived, the sirens are tested the first Wednesday of every month at one o'clock in the afternoon. I came home from work early and waited outside until the siren sounded for a few minutes, then crept in. Sure enough, I found Lucky in the basement resting happily on his blanket.

I doubt if Gellar remembers the conversation, but I do. After Scott excused himself from the table and the other people enjoying their chicken told me next time I was at a table with E. Scott Geller not to argue with him, I thought about the training with Lucky. Lucky wanted a positive reward: comfort and a treat. Lucky also wanted a negative reward, to avoid the loud siren. If he hadn't cared about either of those things, I never would have been able to train him to go to the basement. I'll go more into the differences between rewards and punishments, and positive and negative in the context of rewards and punishments a little later.

I thought about how exactly my experiences with Lucky could help me at my job (other than annoying famous guys with brains ten times as large as mine). Like Lucky, employees have natural behaviors. To train, and indeed, educate those employees (Geller is correct, education is the way to go; I still argue that there is room for pure psychological training, though) I had to observe them and learn what each employee's and each employee group's natural behaviors were. In this way, I could try to enhance those pre-existing behaviors.

I started by experimenting on Lucky. He liked to pick his forepaws up and put them on the cupboard doors, and he liked to reach a forepaw out to me when I was going to give him a treat (he was right pawed). I used these pre-existing behaviors and taught Lucky how to sit up and beg, and how to shake on command. My friends were amazed at this crazy cat who sat up on his haunches on the command of "Prairie Cat" and would reach out a paw to shake like a dog on the command of "High Five." I won more beers betting them that the stories of Lucky were true than I ever won on picking football games. It was great.

It was time to try my technique at work. Employees, I thought, really aren't that different at their core than cats, dogs, camels, cows, or Caledonian crows (holy buckets are crows smart). All animals basically do what will get them a desired reward quickest and with the least effort. As humans, we've got a little more wiring in the control center and a more powerful CPU, but our behavior isn't that different from any other animal.

I decided to experiment with the maintenance crew first. I was the safety nerd for a shoe manufacturer at the time, and the maintenance crew was always tearing machines down to work on them. I spent a half day with two of the more friendly guys in blue shirts, admiring their name tags (I've always had this urge to work in a shirt that has a patch with my name sewn on it) and watching how they went through their lockout/tagout process. The lead maintenance employee pulled out a checklist, and the two men worked through it, step-by-step. I asked, "Doesn't that slow you down?"

The lead nodded and said, "Yeah, but you'll yell at us if we don't do it that way."

I shrugged. He had a somewhat fair point; I wouldn't have yelled at them, I'm not that kind of guy, but I would have asked them a lot of questions that they probably couldn't have answered and that would have slowed them down even more. They continued through their checklist, and when the machine was locked out, they opened up their tool boxes. The tool boxes were jumbles of wrenches, pliers, hammers, small calendars with pictures of topless women, and other interesting stuff. They tore the machine down, fixed the item in question, and put the machine back together again.

It was time to remove the lockout devices. The two employees thought that the really dangerous part of the job was over, and as such they did not follow a checklist when putting the machine back together. When they were done, I asked, "Why didn't you use a checklist to restart everything?"

The lead rolled his eyes at me, and I could almost read his mind. "Ain't gonna cut my finger off now, is it?"

"Probably not. But how do you know you did everything right?"

Now came "The Lecture." If you work in safety, you know The Lecture; you've received it a thousand times if you've received it once. It starts with *I've done this job for twenty-six years* and ends with a speech about how safety people have no idea how to actually do the jobs that keep the business running. I conceded the point. He had done the job since before I could ride a bicycle and I had no clue how to do what he did. Admitting those things to an employee is a great way to A) gain their respect, and B) confuse the living apples out of them. He wasn't expecting me to admit that he was right. By the end of the conversation that followed, he had taught me how to take the machine guards off properly and then put them back on.

I thanked him, told him a dirty joke (another great but risky way to gain confidence and make them wonder just what kind of safety person it was management had hired), and headed back to my office. I analyzed the behaviors I had just witnessed. The employee:

1   Used a checklist because I was watching.
2   Did not use a checklist when he perceived the risk was gone, even though I was still watching.
3   Wanted to show me what he did all day and why it added value to the facility.

My conclusions were grand. If the employee felt like there was *accountability* for his behaviors, he would modify those behaviors on his own. The employee needed to be *educated* about risk, and why something posed a risk. He had never considered that improperly removing lockout could cause an incident. Finally, the employee viewed me, the safety manager, as someone who should see how well he did his job, and know how important that job was. By no means am I the first person to discover or write about these ideas, but when I discovered them they opened up my world.

The next thing I knew I was watching all kinds of people do their jobs, and learning to do those jobs myself. Years later I would learn that what I was doing was called "Gemba walks." I've devoted the next chapter of the book to them, and I will mention them a lot throughout the rest of the book.

> *Gemba walks*
> I will go into detail on Gemba walks later. For now, just understand the basic concept that a Gemba walk is going around learning how people do their jobs and why they do them the way they do (Womack, 2011).

Now my challenge was to change the maintenance lead's behavior. I decided to start with his risk perception. The next time I trained—wait, educated (thanks, Scott Geller!) the maintenance crew on lockout/tagout, I tried something different. I talked a lot about the mechanical process of an amputation.

I didn't do it in an overly graphic, gross-out fashion. I actually can't stand safety training that does that; it's like smacking someone upside the head and yelling *"BE SAFE!"* in their face. Instead, I talked about the way that a nip point grabs the flesh and pulls it in as if it were a leather hide being prepared for cutting. The maintenance guys, who could relate really well to mechanical processes, started to share some stories. Pretty soon, we were talking about a machine in the plant that I liked to call "the rotating wheel of death" that had a series of curved blades designed to chunk out extra material from soft rubber parts. The guys (it was an all-male crowd), shaped their hands like the blade and motioned it into their arms, picturing what it would be like to experience an unexpected release of energy. Their risk perception, because of taking the time to educate them, was now opening the window for me to train them in a more purely psychological way.

I had built a board that had every kind of switch, valve, etc. that existed in the plant and needed to be locked out. I knew that these men did not think I fully understood what they did, and they were right. I also recognized that these men had all been locking machines out for a long time. I asked *them* to educate *me*. One by one, each of them came to the front of the room and locked out everything on the board. Soon, it was a lockout rodeo and they were timing each other to see who was the fastest. I asked if I could give it a try.

They laughed and told me to go for it. Little did they know that I had spent about two hours the night before learning how to use all the lockout devices on all the switches and valves. I rocked the lockout board like Animal from *The Muppets* rocks a drum set. I didn't have the fastest time, but I impressed the men in blue shirts. They were ready to listen more and to talk to me more about their jobs, the challenges of those jobs, and safety in general.

I told them Mamma didn't raise no fool, and they now had enough evidence to believe me. The next thing I told them was that it was really, *really* important to use a checklist to lockout a machine and then use another checklist to remove lockout properly. They nodded. I asked them, "How can I make it easy for you to use a checklist to do this? And will you actually use it?"

They started running the honesty calculus formula through their heads. *What is the most honest answer I can give without creating too much work for myself or getting myself into trouble?* We all know that formula. We all use that formula; it's human nature. I use it every time my wife asks me if I need to get any chores done around the house before leaving for a hunting or fishing weekend. I stopped them before any answers were given, sat on the front table and said, "Look guys, all I want to do is keep you working and going home safe. I'm not here to make more work for you, and I know that the quicker you do your job the easier life is for all of us. The honest answer is the right answer." *The honest answer is the right answer…* I've shared that phrase with countless employees, and I try to live by it (except when I think sweeping the garage will make me late to the cabin). The honest answer is the right answer.

They told me that they didn't want to carry paper around. That it was a hassle to find the checklists, and that they had the steps memorized anyway. In the end, we developed a series of small, laminated pages of instructions that were taped inside the tool box lids next to the calendars featuring topless women. I used the existing behaviors to modify future behavior, by working with the employees and rewarding an honest answer with ideas for how to form a solution around their needs. I had no guarantee they would actually use the checklists, but they would be much more likely to do so than they had been in the past.

That's as far as the safety professional can go sometimes… Leading the horse to water, giving it a straw, and explaining the value of hydration. Now and then, we need to walk away and just let the horse decide whether or not to drink, simply because there isn't anything else we can do.

---

*Paper sucks*

I have yet to meet someone who is thrilled to get a stack of papers to read. People will take all kinds of electronic files and set them aside to read over their Christmas break or on an airplane, but the papers get thrown away. Avoid paper information and emphasize electronic knowledge-sharing.

A few years later, after Lucky had moved out, I lived with two other cats, Mable and Edgrr. The cats were two years apart in age, but had come from the same farm. Mable was, in fact, Edgrr's aunt. They had a bad habit of jumping onto the counters in the kitchen and trying to steal food or lay on the section of counter above the dishwasher because it was warm after the dishwasher finished a cycle. The problem was that cat hair was getting into our food and we had paw prints on the counters, both of which grossed out company. And cats just don't belong on kitchen counters.

We tried electrified mats, but the cats listened for the hum that buzzed when the mat was on, then limited their jumping to when it was not humming. We tried putting double-sided tape all over the counters, but the cats could tip-toe around it, and when covered in tape the counters were useless to us. Come to think of it, that was a period of time when I never spilled a beer, though. It didn't matter what we did, we could not keep the cats off the counters.

At this same time, my wife decided that it would be a great idea to teach the cats to use a people toilet. Over the course of many days, we moved the litter box closer and closer to the toilet, then raised it up a little bit each day by stacking books under it; finally, we moved the litter box onto the toilet. I missed my toilet, and was happy the day we took the litter box away and replaced it with a shallow tray because it meant training was almost complete. The idea was that when the cats were used to the tray, it would be replaced with one that had a hole in the center. The holes would get larger as the trays were changed out. Within a few weeks, we thought we had the cats fully toilet trained.

The same day that Mable got into a bowl of tuna salad on the kitchen counter, we discovered that Edgrr had gone on a litter—well, toilet—strike. The carpet in a mostly unused part of the house was in bad shape. The tuna was catified. It was not a good day.

Two household changes resulted from the bad day: First, we gave Edgrr a litter box back. Second, we ordered an indoor electric collar system. We could not fight the natural behavior of the two individual cats, only modify it a certain amount from the baseline.

It is tempting to say that if I could only have sat down with the cats and explained to them the reasoning behind what we were asking, they might have complied. The conversation would go something like this (you have to picture me and the cats sitting around the kitchen table for this to really be effective)…

Me: Now, kitties, Melissa and I have called this family meeting so we can discuss some of the issues that have been going on.

Edgrr: Grr!

Mable: Rowr.

Me: Great. I'm glad that you're ready to talk. Now, Edgrr, let's start with you. We want you to give the toilet a try for a week or two. Then, we can talk about it again. It's really important that you try it. Litter boxes don't smell good, and you get clay all over the carpet. I know you love your litter box, but I'm asking you to just try for a little while.

Edgrr: Grr!

Me: Thanks, I promise that I will give you a treat each time you go. I know how much you love your tuna snaps.

Edgrr: Grr!

Me: You're very welcome. Now, Mable. We need to talk about the kitchen counters.

Mable: Rowr.

Me: I know, I know. There's good smelling stuff up there. But here's the deal… I know this might be hard for you to accept, but that's people food. And you just aren't people.

Mable (shocked): Rowr?

Me: Nope. We give you plenty of cat food. Please remember what the vet said about your weight and the danger of feline diabetes.

Mable: Rowr.

Me: Okay. Thanks, kitties. I knew you'd understand reason.

Animals do not change behavior easily. It took building an electronic barrier around the kitchen to change behaviors in one instance, and in the other I simply gave up. The lesson: If an employee is behaving in an unsafe manner, remember that it is oftentimes easier and more effective in the long term to change the physical make-up of the job than it is to change the animal's behavior. Reasoning, coaching, behavioral observations… they are not an adequate substitute for creating an environment in which the subject can work according to their natural behaviors.

Alright, a few of you are freaking out right here. I'm calling humans an *animal* again? I apply the term to humans not to degrade, but to be realistic. As discussed earlier in the book, we are a species of animal, not very different from other species. In fact, most sources I found indicate humans and cats share about 90 percent of the same DNA (Pontius et.al., n.d.).

We humans have some basic animal motivations, and we have some unique individual behaviors. Mess with your spouse and play the same Beatles song every time you make chicken fried steak for dinner. After doing this about ten times, play the Beatles song but cook goulash. They won't know why, but they will have expected chicken, because they have learned that that particular Beatles song means chicken fried steak is on the way, just like Pavlov's dogs learned that banging a gong meant dog food was on the way (Cherry, n.d.). As safety professionals, we have to understand and accept the fact that there are only a few things that set us apart from other animals. Those differences are important, but we still breath air, reproduce, eat, and someday die.

How does all of this relate to the typical behavior-based safety system? In a typical system, one employee makes an observation of another employee, then offers coaching or kudos. Remember when I talked about Lucky earlier, and mentioned a positive reward and a negative reward? In psychology there are two types of rewards: positive and negative. There are also two types of punishments: positive and negative. When most psychology students hear this, there is a *Whoa,*

*whoa, whoa!* moment. If something is negative, how is it a reward? If something is positive, how is it a punishment? Think of positive and negative in a different sense; instead of good and bad, positive means adding a condition, and negative means taking a condition away (Prince, 2013).

In a behavior-based system, the kudos are a positive reward. An "Atta boy!" is given to the subject. If an employee is observed enough times and they are judged to work safely, they are not observed any more. Taking the observations away is a negative reward, the same way that removing the sound of the siren by moving to the basement was a negative reward for Lucky. The coaching that one receives in a behavior-based system is a positive punishment. An example of a negative punishment might be taking away a machine operator's radio if it is causing distraction to the machine operator. Behavior-based systems assume that through rewards and punishments, behaviors will be modified or reinforced.

If an employee is already behaving in a safe manner, little reinforcement is necessary. I am a believer in a simple "Thanks" when someone is obviously putting in the effort to work safely when there are easier options. If an employee is behaving in an unsafe manner, the way to truly change the behavior is to make the punishment consistent and unpalatable. We cannot use shock collars on people (thankfully in some cases and unfortunately in others), and we cannot observe an employee 100 percent of the time. My experiences with the cats taught me this: When an undesirable behavior exists, find out why it exists, and modify *the job itself* or *the work environment* to be different.

BBS can add value when used a bit differently than most of the people who sell it advise. BBS believers like to the talk about the "ABC" of behavior, where "A" is the antecedent, or thing that leads up to the behavior, "B" is the behavior itself, and "C" is the consequence. If you can identify something that causes a behavior (the antecedent), but your system teaches users to simply react in a manner that attempts to change the behavior, your user has just witnessed a root cause and ignored it. That's bad safety.

Rather than relying on observations, checklists, etc. to try and "catch" good and bad behaviors, a better investment is using a behavioral observation to guide where to focus time and monetary resources to improve workplace engineering and design. Behavior can be modified only to a certain extent and the employee may revert back to their baseline when not being observed; but the design of the workplace can be greatly and permanently modified with consistently predictable results.

# 9   Gemba walk this way

Nearly all safety professionals have sat through an adult learning course and were taught this basic principle: Adult learning is most effective when it is experiential (Merriam & Brockett, 2007). In my experience, many safety nerds tend to follow this axiom well for others, but forget to apply it to themselves. Most of us can think of at least one peer who writes safety programs that apply to tasks they do not understand; it might be a safety professional who writes a ladder safety program but has never actually used an extension ladder, or who writes a forklift safety program, but has never actually driven a forklift. I spent some time in the Army as a combat engineer, and I shudder to think of what our instructions for complicated explosive demolitions tasks would have looked like had they been written by a physics expert who had never actually blown something to bits with C4.

*Expectations of knowledge*
Safety professionals are often relied on to be unrealistically knowledgeable. Some of the best advice I ever received was to never stop learning, fall deeply in love with the phrase "I don't know, but I will find out for you," and to always figure out how to find answers to questions.

Relying on safety professionals to provide technical guidance on items that they are experientially ignorant of is unwise and dangerous. How can a person provide proper technical guidance if they have never experienced the activity? For that matter, when employees realize that the safety guidance they are asked to follow was written by someone who obviously has not experienced the activity in question, how are we supposed to realistically expect those employees to follow the guidance? Imagine sending your teenage son or daughter to a driving course instructed by a person who has vast safety expertise but has never driven a car... Would you have true confidence in that instructor, or would you rather pick a driving course taught by someone who has no safety management training but has been a licensed driver for twenty years with a good record?

Most Americans remember Chuck Yeager as a gifted test pilot and the first human confirmed to break the sound barrier. What a lot of safety professionals may not know about General Yeager is that he began his military career as an enlisted mechanic, and was made a "Flying Sergeant" when the Army Air Corps needed more pilots for service during World War 2. In his autobiography, General Yeager recounts multiple instances where the knowledge he gained as a mechanic, building, repairing, and diagnosing mechanical issues with aircraft, saved his life as a pilot in combat and when flying experimental aircraft (Yeager, 1986). Yeager's experiences can be used as an analogy to the knowledge a safety professional can gain during a Gemba walk. By taking the time to understand a knowledge set outside of his immediate area of concern (piloting the aircraft), Yeager was able to apply a unique perspective to the challenge at hand (quickly diagnosing and dealing with an in-flight emergency) in such a manner as to provide for the most favorable outcome. Had General Yeager not understood the mechanics of how his aircraft worked, he would not have had the skillset to diagnose problems and understand the capabilities of his aircraft. In a similar fashion, a safety professional can gain knowledge outside of her or his immediate area of concern to apply unique perspectives to problem solving when EHS challenges arise by engaging in the practice of Gemba walks.

As safety professionals, we may not be flying experimental aircraft; but the lives and livelihoods of employees, the wellbeing of our communities, and the protection of the environment are unquestionably affected by our decisions in very real ways. I am not advocating that all safety professionals put their career on hiatus and spend the next three years roofing, driving forklifts, assembling widgets, or logging. Instead, I propose that the profession embraces the concept of Gemba walks to achieve experiential learning, and that classroom time spent learning trade work otherwise unrelated to safety be accepted as continuing education credit.

As opposed to management by walking around, a Gemba walk is taking some time to learn and understand the work being done by employees (Womack, 2011). As safety professionals, we need to take this concept one step further, and dedicate scheduled time each month to perform work with employees, side-by-side; a portion of one's continuing education time each year should be dedicated to learning details about a technical skill related to the place of employment (such as becoming a certified forklift operator or attending a trade-related class).

*The American Society of Safety Engineers*
ASSE is over a century old, and is the primary association for safety professionals in the United States. Professional Safety is ASSE's monthly peer-reviewed publication. If you are thinking of becoming a safety professional, visit www.asse.org for information about the profession. If you are already a safety professional and are not a member of ASSE, please visit their website for information on the advantages of membership.

Please note that the American Society of Safety Engineers is, as of the printing of this edition, voting on a possible name change. If you are an earthling from the future reading this, the organization may in your time be called the American Society of Safety Professionals. Also, if you are a future earthling, I'm really sorry we screwed the planet up so badly. My kid became president and fixed it, right?

Not only will working with line employees to learn their jobs improve your understanding of how the work in production areas is actually done, but it will result in relationship building and contribute to direct, unfiltered feedback from employees. In their January 2015 *Professional Safety* article titled "Safety Conversations," Carrillo and Samuels point out the importance of this type of feedback, as well as the gains that can be had through meaningful conversations with employees. In my personal experience, once an employee witnesses me do something like stop touring a grain handling facility and begin shoveling corn with them, or sitting down next to them and actually trying to assemble a fiber optic component, the employee's attitude toward me (and indeed toward safety as a whole) becomes less guarded, and employees begin to provide substantive feedback and observations. Employees who I have worked with in this manner will now flag me down as I pass through a department and show me something they feel could be a safety issue. In other words, I'm no longer that odd safety guy who wanders around the facility; instead, I am more of a team member and a friendly acquaintance.

Once a safety professional has taken time to show employees that the work those employees do is not "beneath" the safety professional, the EHS Department can create more realistic policies and procedures that align with the safety professional's hands-on experience performing the work tasks.

---

*Chuck Yeager on gut feelings*
In his biography, Chuck Yeager discusses a day during World War 2 when he had a feeling of dread in the pit of his stomach about flying. His mission was called off due to weather shortly before takeoff, and Yeager believes that had he flown that day, he wouldn't have made it back to base. In my career, I have seen enough similar examples to tell people to trust their gut. If something feels wrong, something is probably wrong.

---

Let's visit General Yeager again for some more insight. One of the reasons he was a superior test pilot was his ability to relate to the men who built and maintained the experimental aircraft Yeager piloted. Yeager worked closely with these crews before ever leaving the ground in order to have a fuller understanding of the true capabilities of the aircraft he was operating. This understanding saved his life in multiple situations (Yeager, 1986). In the same way, a safety professional can create policies that are more reflective of employees' true needs and capabilities once the safety professional has learned what it feels like to hold the parts, turn the screws, or bounce over roads in the driver's seat. The result? Lives saved.

Getting "in the trenches" with employees also allows safety professionals to perform more effective incident investigations. During an investigation, the investigator can *listen* to an employee describe why the guard on a cutting tool was removed, or the investigator can walk through the performance of a task with the employee and *experience* why the guard on a cutting tool has been removed. Anyone can look through policies, talk with employees, take a few pictures of a removed guard, and come to a conclusion that the employees just need to do the job the way *we tell them to*. It takes something more to go to the workstation and have an open enough mind to experience why the employee removed the guard to do the job *the way they felt they need to do it*. By learning the true reasons why employees perform their jobs in an unsafe manner, the safety professional can get to the true root cause of an incident to make engineering improvements.

Dedicating time and putting forth effort to work side-by-side with employees also brings with it the long-lasting benefits of gained credibility and respect. Early in my tenure in a manufacturing facility, the facility maintenance technician was assigned a terribly loathed task: cleaning out an in-floor harpoon system that pushed oil and metal shavings out of the machining department through a shallow trench and into a waste tank. There was a deadline for the task, because the company was moving out of the building, and the entity that bought the facility was going to do their final inspection in less than a week. While many engineering and accounting managers had promised to contribute an employee or two to help out, none of the managers delivered on the day the work had to be done. I heard about the situation from the maintenance technician, and immediately left work to go home and change into clothes that could get filthy. I spent the next nine or so hours performing a job that absolutely nobody in the facility ever wanted to do. By the end of the day, I was covered in oil and had decided to just throw my clothes away rather than try to clean them. Luckily, there was a locker room with showers at the facility, so I could chuck my dirty clothes in the trash there, take a shower, and change back into clean clothes.

---

*Know what you are getting into*
Don't show up to do a Gemba walk of a dirty job in your dress clothes, it is insulting to the employees who do the job, and shows how clueless you are. A little research into proper dress goes a long way. Also, make sure to ask employees how to do the job safely; they will respect you for it, and might otherwise assume you already know how to do it safely and let you do something stupid that gets you hurt.

---

Word spread that the new safety guy was willing to leave his "cushy" cube and get dirty. Employees hadn't learned yet how much I hate being stuck at a desk, but they did know I labored through a long, hard day, that afterward I needed to play catch-up with emails and other work, and that I had helped turn a two-day miserable, lonely task into a one-day only somewhat horrible and much less lonely task. From that day forward, the entire maintenance department was one of the strongest groups of safety advocates I've ever seen. When I wanted or needed to make changes in safety procedures, I sat down with them and they gave me frank, blunt, sometimes quite colorful feedback in an open and respectful manner (they informed me that they were trained in Deming's principle of calling it like it is). In the end, they understood that if a new safety rule meant they had to make some changes to the way they did things, they were more accepting because of the relationship we had built.

Even though most of the safety professionals I have gotten to know over the years do not feel like they are really a member of the management team, many employees view us as a part of "management" because we make rules and expect people to follow those rules. If safety professionals can accept that we are perceived as part of management, then we can also understand why our leadership style matters. If we truly want employees to buy into our programs, we have to follow some basic concepts of effective leadership. The following actions are suggested by the American Management Association to help avoid destroying employee morale (Schaefer, 2014). Here, I have applied them to safety management.

1   **Form relationships built on trust.** In safety, this means dedicating time at workstations to build relationships with employees, to illustrate through actions that you are interested in *how* they perform their work the way they do, *why* they perform their work the way they do, and that your safety solutions will conform as much as possible to the way they want or need to do their work.

2   **Show them respect.** If a safety professional creates new policies that affect a job without first taking the time to learn how and why people do the job the way they have been doing it, it shows a lack of respect. Have you ever reported to a supervisor who had little or no understanding of your job? How many times did that person set unrealistic rules or expectations for you? In the same way you wish to be respected by those who do not understand your job, workers should never be made to feel like they are seen as inferior to managers or more highly skilled and technically trained employees. By taking hands-on time to learn the basics of how jobs are done when it is time to create safety policies, the safety professional is showing employees that she respects the efforts made toward completing their production goals.

3   **Nurture creativity.** When taking time to work side-by-side with employees and discussing how and why jobs are performed a certain way, employees will often come up with novel ideas for safety improvements. If the safety professional responds with a display of genuine gratitude for the possible discovery of a safer way to do a job, the result is a stronger relationship. Once a trusting, emotionally safe relationship is built and employees know that the safety professional respects them, they should then be more open to offering their ideas for improving workplace safety.

4   **Build effective teams.** Team building can be challenging, but it becomes easier once the safety professional has taken time to become "literate" regarding the various job functions in the workplace. Need to form a team to engineer a lifting solution in a tool and die shop? If the safety professional has taken the time to learn some of the common skill sets around the facility, she can cherry-pick team members who are likely to contribute.

5   **Make it real.** Show genuine interest in learning how and why people do their jobs the way they do, and they will feel respected. If the safety professional spends time learning jobs, but never expresses interest or curiosity about that job to employees, employees will become less likely to provide open and honest feedback in return. People like to talk about what they do for a living and why it is important, so ask a lot of questions and listen more than speak.

Some jobs take years of training in order to achieve proficiency. If the safety professional approaches the employees performing those jobs and announces they'd like to spend a day or two learning the job, it can be viewed as flippant or as not fully appreciating the skill that goes into the job (Bob, I'd like you to spend the next hour teaching me how to assemble this nuclear reactor; if you can learn it after years of schooling and apprenticeship, I can certainly figure out the basics before lunch!). Careful communication is required to show employees that the safety professional does not think she will become an expert at the job, but simply is looking for a basic understanding of how and why the job is done the way it is.

Think because you never took any shop classes in high school and your college major was art appreciation that you don't have the chops to go out and learn the various jobs in your organization? Use Mike Rowe as your inspiration. Most people know who Mike Rowe is, he makes his living learning how to do other people's jobs on TV. If you've ever seen one of Mike's shows, you know that he doesn't pretend to become an expert in the time it takes for him to film the show, but he does a pretty good job at learning (and sharing) what the job is like. Mike comes across as a "working man," the type of blue collar guy who grew up with a wrench in his hand and spent his high school years taking cars apart. Far from it. Mike was a choir geek in high school who became a professional opera singer before moving to TV and eventually hosting shows like *Dirty Jobs* and *Somebody's Gotta Do it*, where he entertains audiences through his adventures taking Gemba walks (Rowe, n.d.). Plan ahead, dress for the work, and don't be afraid to admit that you don't know what you are doing; employees will respect you for it.

The final point to consider is what to do if a job is extremely dangerous. Bureau of Labor Statistics data indicates the fatality rate in the agriculture, forestry, fishing, and hunting sector was 22.2 fatal injuries per 100,000 full-time equivalent workers in 2013. Should the safety professionals and other managers who work in the above industries realistically be expected to perform these risky tasks? Absolutely! Eleanor Roosevelt famously said "It is not fair to ask of others what you are unwilling to do yourself." If a job is so inherently dangerous that you are unwilling to do it, why should anyone else be willing to do it? In these cases, take time to train for the job and to learn safe practices from those who perform the work before joining the job activities. If it is simply impossible or infeasible to perform the tasks (lineman work for instance), then make arrangements to be an observer. A great plan is to join a group of new employees for their basic training period, then spend time on the job. That's right—those employees you are writing safety programs for are the ones who will teach you how to do the work safely. Not only will you learn how and why the job is done the way it is, but you will be given a chance to evaluate new employee training in the most effective way possible. It might be working near rail cars, performing a job at a dangerous height, or working near machinery. After you have gone through the training process, you will either have confidence that new employees are being trained properly, or you will have a new number one item on your to-do list.

---

*Wear PPE*

In the spirit of Eleanor Roosevelt telling us not to ask others to do what we are unwilling to do, I advise all safety professionals out there to wear the personal protective equipment (PPE) they ask employees to wear. It won't take long before you understand why they complain about it so much.

---

When I was a fresh, shiny new safety professional, I was dealing with a safety situation where an employee working with adhesives all day kept a small bottle of acetone at his workstation. I asked him one day what it was for, and he said "To wash my hands." I did a triple-take.

Pointing to the bottle, I said, "You wash your hands with acetone?"

He nodded, having no idea why I was shocked. I immediately gave him instructions to use the sink in the bathroom when he wanted to wash his hands, took the acetone away, and spent an hour in my office shuddering. Three days later, I saw an acetone bottle at the workstation. I asked the employee, "What are you doing with this?"

He shrugged and said, "I wash my hands with it."

---

*Skin exposure to acetone*

Acetone isn't much of a health risk unless it sets on fire while you've got it on your body. Still, it is always best to reduce chemical exposures.

---

We danced the dance one more time. I was frustrated. How was I going to explain to this employee how unsafe it was to coat his hands in a flammable liquid? I walked a lap of the facility to think, returning to his workstation after shuddering and sighing about the situation in only the way that a shiny new graduate can do. I watched him work, waiting for a natural break to stop him and say something... anything that would just make him stop using acetone to wash his hands. As I watched him work, I saw how the adhesive squeezed out of the product he was building and stuck to his hands. After about five cycles, his hands were like two bricks of tar. He turned for his acetone, remembered that I had taken it away, and I saw his face and shoulders melt toward the floor.

I tapped him on the shoulder, now feeling like I had just kicked a puppy, and told him that it looked like it was pretty hard to function with hands that get covered in thick glue all the time. He nodded, and asked for his acetone back. I couldn't ethically give it back to him. I explained this, and we found him a rag he could wipe his hands with. I asked if he could show me how to do his job. I knew that I would never understand his situation until I had lived it. I did the job for about twenty minutes. It didn't take long to realize that a change was needed. I conferred with plant engineers and called the glue company. We weren't able to change the process, but what I learned from the glue company was that there was a great hand cleaner available. It was waterless and took the glue off like a charm. I ordered a huge tub of it, put a bow on it, and hand delivered it to the employee's workstation.

He laughed and thanked me, then pointed to another process. "You know how the people who do that have shoulder problems?" I said I did. He told me, "Do their job like you did mine. You might come up with a way to help them out."

Pretty soon I was doing every job in the factory. The employees looked at me like I was nuts. Here came this safety guy; a guy that, in their eyes, didn't know what it was like to work a long day that became a long week, month, year, and career. What they saw was someone who had gone to school to work in an office rather than getting his hands dirty. And here this person was, asking them to teach him how to do their job. What they never knew (until they decided I was a good egg and asked me out for a beer or six), was that I had worked long factory shifts over the summers while in college. I had served in the military. I crushed a finger unloading rocks once when I was fourteen so badly the insides of my knuckle were on the outside. In short, I knew how to do "real" work.

I think that interns and other safety professionals who have worked with me were thrown a curve in the past when I've made them unload trains, drive forklifts, and do other "odd" tasks. One intern celebrated when she climbed a series of fixed ladders to the top of a grain elevator and conquered her fear of heights, taking a selfie to send to her mom. While an employee who hasn't met me yet will give me odd looks when I grab a shovel or throw on leather work gloves, they really get a kick out of interns doing it.

---

*Working for a TPA*

A TPA is a third party administrator that administrates workers' compensation for self-insured entities. It is great work for safety professionals (especially those who hate being stuck in an office), because it provides an opportunity to learn about a wide variety of workplaces at the locations where work is being done.

---

I worked for a few years as a loss control consultant with a TPA, handling workers' compensation programs for cities around the country. I decided to do a Gemba walk with a street crew, filling in pot holes. When I arrived at six in the morning, they asked me three times if I was really sure I wanted to do this. I told them I didn't get up that early for nothing, and away we went. I spent eight hours walking down streets behind a dump truck, filling a shovel with hot asphalt and packing it into pot holes. After my first couple of scoops, I felt pretty good and thought I was doing a really nice job. Then one of the guys on the crew said, "Hey Teaspoon, give me a hand packing down this crack fill." The rest of the guys laughed.

I asked, "What do you mean by 'Teaspoon'?" after taking a moment to confirm to myself they hadn't been laughing at the phrase "packing down this crack fill."

Another member of the crew told me, "Fill up your shovel." I did as I was told. He then proceeded to fill up his shovel with about three times the amount of hot asphalt. He looked at his shovel and said, "Shovel full." Then he looked my shovel and said, "Teaspoon."

This Gemba walk not only taught me how to do the job in question, but it gave me a great lesson about the culture of the employees. No wonder they were lifting so much and hurting their backs! If they didn't lift an amount that was perceived to be adequate, they were teased by the rest of the crew, who felt like they had to do more to make up for the lazy or weak employee. Smaller shovel fulls would also mean a longer shift with the hot materials. The musculoskeletal injuries they were experiencing were culture-based as much as they were action-based.

There is a ton of value for the safety professional in a Gemba walk. It is important to remember that the employees may not see you as equal to them in strength, willingness to get dirty, or practical ability. By performing a Gemba walk, and honestly putting forth effort to learn what the employees do, the safety professional:

- gains the respect of the employee population;
- learns why employees do the job the way they do it;
- increases the ability to find solutions to the safety issues the employees face;
- receives honest answers from employees, along with demonstration and instruction; and
- becomes more aware of the company culture at the "boots on the ground" level.

Whether the Gemba walk involves sitting at a computer workstation and entering report data, or giving your palms a few "pride blisters," it is the best way to illustrate to employees who you are and why you ask them to do things like wear protective gear or show up to safety training. They do not know your history; as the saying goes, "Lead with your actions."

I love Gemba walks so much that I actually did one at the state fair on a day off. I wasn't in the mood to do some safety work, mind you; I was just in the mood to see what it was like to work at a fair booth. The "Eat Minnesota Turkey" booth was empty, so I sat behind the counter and handed out leaflets, reading a few notes I found behind the counter as people came by.

A very nice older woman stopped at the booth and smiled. I asked her if she knew that Minnesota was the number one turkey producing state. She said that she was aware of that fact.

"Did you know," I asked, "that the average American eats almost eighteen
   pounds of turkey every year?"

"Yes, I know that."

"Wow," I said, "you sure know a lot about turkeys."

"I ought to," she replied with a smile, "this is my booth!"

We had a laugh, I asked her a few questions about turkeys and working at
the state fair. My wife, who had been looking at a few of the craft booths I find
boring returned and retrieved me, asking me not to man an empty booth again.

The use of Gemba walks and experiential learning can reap great rewards
and become an invaluable tool for advancing your safety initiatives. The time
and effort spent with employees on their day-to-day tasks is an investment in
the future of safety at your organization, and in your proficiency as a safety
professional. Your ability to create meaningful, realistic safety programs will
profit, because in the end not only will the employees gain a much greater
appreciation of you, but you will have a much greater appreciation of their daily
challenges.

# 10 Safety management system title match

## Deming versus Heinrich

Bob Knucklemuncher: Welcome to Friday Night Fights, folks! This is an exciting one, we've got two champion heavyweights of management systems going head-to-head. In this corner, wearing the yellow and black warning stripe trunks, Herbert William Heinrich! And in this corner, wearing the red and white rising sun pattern trunks is William Edwards Deming! I'm Bob Knucklemuncher, calling the fight tonight with my cohost, "Glass" Joe Takeda.

Glass Joe: Good to be here tonight, Bob. The two men come out with a respectful handshake, Heinrich looking focused and quiet while Deming seems to be counting the number of steps from his corner to the center of the ring. Now Deming has said something quietly to Heinrich and Heinrich is laughing. Looks like Deming has the edge on the sense of humor.

Bob: There's the bell, and the fight is on! The two men circle each other, neither of them especially graceful. Deming is expertly coached from his corner, with English biologist and statistician Sir Ronald Fisher cajoling him to throw the first jab. But it's Heinrich who swings first, publishing *Industrial Accident Prevention, a Scientific Approach* in 1931. The crowd goes wild, but it's a wide miss.

Glass Joe: That's right, Bob. Looks like he based the book on subjective findings that can't be replicated. The finding that 88 percent of accidents are the employees own fault looked like a big uppercut at first, but Deming easily avoided it. We'll see how the judges score it at the end of the round.

Bob: Now here's Deming with a big overhand of his own, connecting with Heinrich's left eye by using robust statistical study to claim the exact opposite by saying that 85 percent of all failures are the fault of the management system.

---

*Deming and the red bead experiment*
In the red bead experiment, Deming uses a simple analogy of collecting beads randomly into a tray to show that a focus on results instead of improving the process does not increase production. There is a great video of it on YouTube.

Glass Joe: It's a good fight now, with the consultants in the arena divided. Many of them are chanting "B-B-S!" while others are cheering "Red Bead Experiment!" There's the bell and the end of round one. Here comes our ring card, carried the first time tonight by the highly respected and ahead of her time Francis Perkins.

Bob: Francis is not only a hero to women around the world, but an indispensable historical figure for those who advocate safety and one of the undisputed safety champions of the world.

---

*Important women in safety history*
I know that some of you are freaking out because I have Francis Perkins and other historical female figures included here carrying cards between boxing rounds. Please note that I am going for a little irony and am trying to not just make a point about the society we live in, but to educate the readers on some important historical figures they may not otherwise know about.

---

Glass Joe: Bob, I really think she should have been the one to make the ten dollar bill.

Bob: Couldn't agree with you more, Joe. There's the bell, and our fighters are back.

Glass Joe: Heinrich takes another jab from Deming, who makes contact with his proposition that the key to system success is a focus on the process and not the outcomes; but now Heinrich gets in a body shot by dedicating a huge portion of his book to the importance of machine guarding.

Bob: I think he's just helped give birth to the hierarchy of controls with that one, Joe.

Glass Joe: Oh my! Another big miss by Heinrich, though. He just swung for the fences with his proposition that for every 300 minor incidents there is one major injury.

Bob: That was easily blocked by Deming, Joe.

Glass Joe: You really can't use an argument like that against a statistician, Bob. Poor attempt by Heinrich, although many of the consultants in the arena seem to love it.

Bob: With these two strong sets of ideas, they're going to make a lot of money on this fight, Joe. There's the bell and the end of round two. We're going to take a quick commercial break, and then it's back to Friday Night Fights.

---

*This never happened*
Just to make sure I am clear, Deming and Heinrich never debated or otherwise competed against each other in real life. At least not that I know of… I guess it's possible they could have been in the same hockey league or something.

---

Commercial announcer: Got a safety problem? Not sure what to do? Hate taking responsibility for the systems you design and manage? Have I got a solution for you! Just use this simple technique of training the people who investigate your safety incidents to believe that there is a 95 percent chance that the accident is the employee's fault. Human psychology being what it is, once you do this, there is a nearly 100 percent chance that your newly trained investigators will blame *every* accident on employee behavior to create a self-fulfilling prophecy. Call the number on your screen now for details.

Bob: Welcome back to Friday Night Fights. Here is our next round card, carried by the incredible Rose Schneiderman, the self-educated union pioneer and women's rights activist. Rose takes a bow to thunderous cheers from the audience, thankful for their eight hour workday.

Glass Joe: There's the bell, and our two fighters are back at it. Deming comes out aggressive, pounding Heinrich with his System of Profound Knowledge and Fourteen Key Principles.

Bob: Joe, a group of the consultants rooting for Heinrich's corner are booing and lodging a complaint with the judges, claiming that Deming wasn't even focused on safety. The referee has stopped the fight while this gets sorted out.

Glass Joe: This is an odd turn of events, Bob. Could be a TKO called here.

Bob: Okay, it looks like we are going to start the fight again. The referee is indicating that the judges have determined that good safety performance is one of many measures of good overall business management. Therefore, safety should be treated just the same as any other important function or aspect of the company.

Glass Joe: Wow! We just barely heard the bell, and Deming is at it again. He delivers a big blow to the ribs of Heinrich with Plan-Do-Study-Act.

---

*Shewhart*

Walter A. Shewhart was a pioneer of industrial quality control. The Plan-Do-Check cycle that Deming is often credited with was originated by Shewhart.

---

Bob: Great technique. Originally developed by Shewhart when he won the Statistical Quality Control world belt back in 1924. And it has worked brilliantly here tonight, as Heinrich seems to be losing steam.

Glass Joe: Uh-oh, Bob! Looks like Heinrich's corner and several of the spectators are storming the ring! They're calling for a disqualification!

Bob: This just became more of a wrestling match, with both of the boxers now fending off the consultants storming the ring, who've gone wild. The referee is clearing them out now. Many of the consultants seem to have gotten lost, and cannot remember which seats they originally came from.

Glass Joe: Bob, it looks like they're going to call this round off early to let things settle down a bit. For our home audience, we're going to take another break. A wild Friday Night Fights will be back after these words.

Commercial announcer: Hey workers, are you tired of getting blamed for everything bad that happens at your workplace? Try this simple pill,

*Demingway. Demingway* instantly transfers all of your responsibility onto the system and your managers. Once you've gotten the pill into your system, you won't even receive annual reviews! Just look at this happy young lady. Six months ago, she was nearly fired for texting at her workstation but today she is free and clear, all because of *Demingway! Demingway* is available at your local melon stand.

Bob: Welcome back to Friday Night Fights, where we've got a thriller between Herbert Heinrich and Edwards Deming.

Glass Joe: And our third card carrier this evening is rear Admiral "Amazing" Grace Hopper. Without her, we wouldn't be experiencing the technological revolution of today.

Bob: That's right, Joe. Admiral Hopper was instrumental to the invention of modern computers and computing language. Did you know that she coined the term "bug" for a computer malfunction after finding a moth in a Mark II computer relay back in 1947?

Glass Joe: Not only am I aware of it, I've seen the actual moth at the Smithsonian.

Bob: My goodness, Joe! What a sight that must be.

Glass Joe: It looks just like one my cat ate the other night. And oh, my… One of the consultants in the crowd must have had a few too many, as he has tried to steal the card from Admiral Hopper.

Bob: Bad idea. The one hundred and five pound Hopper not only avoided losing the card, but has knocked out three of the consultant's teeth.

Glass Joe: There's the bell, and the two fighters are in the center of the ring, circling each other again.

---

*Competitive ideas*

Fred Manuele and Dan Petersen have published wonderful works to advance the science of safety. I met the late Mr. Petersen while attending graduate school, and was impressed with his ideas regarding perception surveys; I consider Fred Manuele one of the true legends of safety. Neither of these men would hit someone with a chair; safety scholars in general are some of the most generous, gracious people I've ever met. I am a fan of both Mr. Petersen and Mr. Manuele, and admire their works greatly.

---

Bob: Who's that! Things just got crazy, as it looks like Dan Petersen has entered the ring to back up Heinrich with some attempts at modified research. He's delivering quite a few blows to Deming, who is countering with the Seven Deadly Diseases.

Glass Joe: This match is out of control again, as Fred Manuele is now in the ring, hitting Heinrich over the head with a folding chair!

Bob: The judges are banging wildly on the bell as the action in the ring gets hard to follow, with the consultants in Heinrich's corner now being overrun by consultants from Deming's corner who have seen the open niche for statistics-based canned safety programs!

Glass Joe: There are now extra referees pouring into the ring trying to sort this thing out. We've got to go to commercial or the FCC is going to fine us more for the language being used in the ring than OSHA would if we were to electrocute one of our contractors.

Commercial announcer: In the mood for a bat-claw crazy safety program? Try bingo! That's right, take everyone's life and make it into a game that assumes employees consciously make an effort to get hurt unless they are rewarded by management for the natural instinct of avoiding pain and possible death. It's quick, fun, and easy! Best of all, management ALWAYS wins! Has it been 29 days of employees refusing to report an injury so they could earn an extra bingo card? Quick, make sure that the administrative support specialist who asked for an ergonomic evaluation goes to the chiropractor. That was close! Now that you've had one recordable, you don't have to pay out the bingo prize, and you're not the bad guy... The administrative support specialist is! Bingo!

Bob: We're back to Friday Night Fights, and the ring has been cleared. Here's the decision by the judges.

Judge: The judges have found that Heinrich's work, while pioneering and important for its time, is not valid for today's environment. Additionally, we have determined that application of good management practices to safety improves the workplace as shown by examples in industry and research. The split decision goes to Deming.

Glass Joe: Listen to the crowd! After the mass ejections and arrests at the end of the fight, there are only six spectators left in the arena, but they are going wild!

There are several books out there about both Deming and Heinrich, and by no means am I the first person to advocate the application of Deming's principles to safety. There is a real challenge in safety to find the balance between holding the system accountable and holding employees accountable. One of the best recommendations I have to exploring this concept is the book *Just Culture: Balancing Safety and Accountability* by the wonderful safety researcher Sidney Dekker. Many of Dekker's concepts are reflected in this work, as my experiences have drawn me to many of the same conclusions Mr. Dekker has proposed. Both systems and individual accountability are important and a well-balanced safety program strikes a balance between the two. That balance may at times tip more to one side than the other, and in general I am a firm believer that it should always be tipped farther in the direction of management systems.

Below is my brief take on how to apply Deming's principles to workplace safety. Please remember that to really put this into practice, you should take on a serious study of Deming, beyond the information presented here.

Deming's fourteen principles (and ways to apply them to safety):

1 Create constancy of purpose toward improvement of product and service, with the aim to become competitive, to stay in business and to provide jobs.
   • How to apply this to safety: The constancy of purpose should be directed toward improvement of leading indicators, with the understanding

that good safety contributes to a healthy profit. Employees, managers, and shareholders are the safety professional's customers and we need to focus our efforts accordingly.

2    Adopt the new philosophy. We are in a new economic age. Western management must awaken to the challenge, must learn their responsibilities, and take on leadership for change.
   - How to apply this to safety: Although Deming's philosophy isn't new to industry anymore, it seems to be a new idea to quite a few safety managers. There are those among us who are going to read this and think it is old news, and those readers are correct; it is time for the rest of our profession to acquire the bravery required to shift away from behavioral safety and lagging indicators into a systems approach that uses leading indicators as our key performance goals.

3    Cease dependence on inspection to achieve quality. Eliminate the need for massive inspection by building quality into the product in the first place.
   - How to apply this to safety: This can be distilled into the concept of prevention through design. By building safety into the process in the first place, we are less reliant on the need to constantly conduct safety inspections. The higher we move in the hierarchy of controls, the more control we have and the less we need to rely on inspections of possible variables.

4    End the practice of awarding business on the basis of a price tag. Instead, minimize total cost. Move towards a single supplier for any one item, on a long-term relationship of loyalty and trust.
   - How to apply this to safety: Loyalty and trust between you and your safety suppliers creates a relationship where the supplier understands your unique challenges and needs, and is willing to help create the best designs for your systems in order to retain your business. Corporately, apply this to the use of preferred vendors (something that most companies already do).

5    Improve constantly and forever the system of production and service, to improve quality and productivity, and thus constantly decrease costs.
   - How to apply this to safety: Strive for continual improvement of safety performance. Deming teaches us that the focus should be on the process, not the outcomes; therefore, the improvement should be in leading indicators. As we improve our leading indicators, the lagging indicators will likely follow and the EHS department will quietly contribute to the company's bottom line.

6    Institute training on the job.
   - How this applies to safety: This one seems obvious to me… How can a person safely do their job if they haven't received training? Be sure that job training focuses on safety, the safety portion of job training is well-documented, and that trainees are encouraged to ask questions. Apply this to yourself by using the practice of Gemba walks.

7   Institute leadership. The aim of supervision should be to help people and machines and gadgets do a better job.
- • How to apply this to safety: It must be drilled into the heads of supervisors that good safety is part of good management. By focusing on the process instead of the results, a supervisor can identify where there are weaknesses (safety deficiencies) and can work to address them in an appropriate manner.

8   Drive out fear, so that everyone may work effectively for the company.
- • How to apply this to safety: If there is a focus on zero injuries, workers and supervisors will have an inherent fear of reporting incidents. Accept the fact that zero injuries is only possible over the long term if there is zero risk. Because zero risk is not realistic, a goal of zero injuries is therefore not realistic either. Any focus on lagging indicators feeds fear, whereas a focus that is totally pinpointed on leading indicators and rewarding honesty drives out fear.

9   Break down barriers between departments.
- • How to apply this to safety: Remember that good safety is a part of good management. Good accounting is part of good management; so goes it with sales, quality, customer service, research, and a myriad of other practices. If the safety professional can communicate and cooperate effectively and often with other departments, the overall safety systems will improve because the safety system will not just exert influence upon but receive influence from other departments. For example, it is much easier to determine the return on investment of a safety improvement if you've got a friend in accounting who can help you with the math.

10  Eliminate slogans, exhortations, and targets for the work force asking for zero defects and new levels of productivity. Such exhortations only create adversarial relationships, as the bulk of the causes of low quality and low productivity belong to the system and thus lie beyond the power of the work force.
- • How this applies to safety: My personal slogan is "I hate slogans!" Asking for zero injuries is simply not realistic. Instead, work cooperatively with employees to identify the most important leading indicators of safety and create programs to nurture those leading indicators. By showing workers that you are interested not in counting days between amputations but in making their work environment truly safe, you are harnessing the safety power of the workforce. As we will examine later, banners that read "Safety is #1!" are a lie that employees do not respect.

11  Remove barriers that rob the hourly worker of his right to pride of workmanship. The responsibility of supervisors must be changed from sheer numbers to quality.
- • How to apply this to safety: Actively solicit input from employees regarding safety systems. Forget about flavor of the month initiatives, and instead focus on reinforcing the idea that doing the job safely is

part of doing the job well every hour of every day. A lack of incidents does not equal safety, and employees should be rewarded for working in a safe manner rather than simply getting lucky they weren't killed. Think of it like this: I know someone who used to drink and drive quite a bit. When he would leave a party, people would tell him to call when he got home so everyone knew he made it back safely. Never did he make it home safely, but he *did* make it home alive each time. Using lagging indicators, he was a resounding success; whereas the friend of mine who through no fault of his own was killed when he was rear-ended by a semi was a complete failure. Assuming that employees are safe because there are no injuries reported is no different than stating a drunk driver made it home safe because he didn't die or kill anyone else on the trip.

12  Remove barriers that rob people in management and in engineering of their right to pride of workmanship. This means, *inter alia*, abolishment of the annual or merit rating and of management by objectives.

- How to apply this to safety: I have a bit of a tough time with this one. As much as I would love to never have another annual review (giving or receiving), I feel the need to modify this. Maybe my views will evolve and in ten years I'll write a book about how I was wrong on this point, who knows. The best application of this point is to focus on continuous evaluation, especially with Millenials, who tend to love feedback. By the time the annual review comes around, you've had weekly or monthly meetings to check in and make sure everything is copacetic; the annual review, then, becomes a zero stress event where you can collaborate on a plan for success in the coming year.

13  Institute a vigorous program of education and self-improvement.

- How to apply this to safety: Don't you just love annual safety training? I think it is everyone's favorite activity of the year. Wait a minute, no I don't. Safety training generally is not popular. Employees see the same information presented the same way year after year. I don't blame safety professionals, because how many different ways can you present the legally required information? Safety training can have usefulness beyond meeting legal requirements when it helps employees grow and become better at something. This can be accomplished by including home safety information in training, offering CPR and first aid courses, and by educating employees (a la Scott Geller) rather than simply training them. Also, solicit input from employees on safety topics they want to hear more about.

14  Put everybody in the company to work to accomplish the transformation. The transformation is everybody's job.

- How to apply this to safety: First, don't do it with a slogan. There is, as Deming points out, a massive training requirement here. Everyone needs to understand the goal: Better financial performance through a superior safety program. Then they need to understand what a superior safety program is. That means explicitly stating to employees that it does not mean zero injuries, but instead means better performance related to leading indicators. Be careful how you present this or it will sound something like, "We know that some of you are going to get hurt, and that's OK as long as we make money." Instead, it should sound something like, "There is always going to be the chance of an incident, because there is always going to be some level of risk. If we do a better job performing certain key leading indicators, there will be fewer safety incidents, the incidents that happen will be less serious, and the company will make more money."

Deming has the edge over Heinrich, and good safety is good management. Use the concepts of Deming wisely to allow a safety program to flourish. Not only can you drive up your performance numbers related to leading indicators, but managers outside of the safety department who have been trained on Deming's principles will respect your propositions when they understand how they are rooted in the same ideas they use for quality and production.

# 11  Written in blood
## Bhopal India

When I was a kid, I enjoyed watching the news with my parents at night and Walter Cronkite was my hero. I was always very curious about what was going on in the world and why, and for some reason news about fires, crashes, and rescues always peaked my interest (I was doomed to be a safety nerd when I grew up). I remember the night in early December 1984 when I first heard about the Bhopal tragedy. My nine-year-old mind had a hard time wrapping itself around the number of people killed and injured, or how such a thing could ever happen. It was one of the seminal events that led me to my chosen career path.

---

*Unlucky number Sevin*
Sevin and other pesticides made with carbaryl are still popular in the United States (Metcalf, 2000). It is important to remember that even though this chemical is hazardous to create, it has important economic and food supply benefits.

---

Sometime during the night of December 2 or early morning hours of December 3, 1984, water was introduced into a tank that contained 42 tons of methyl isocyanate (MIC). The MIC was used at a Union Carbide facility to produce an insecticide called carbaryl, sold under the name "Sevin." The water and MIC reaction vented from the tank, creating a poisonous cloud that rolled through the shanty town surrounding the plant (Varma & Varma, 2005).

There were likely a handful of toxic gases produced and released by the reaction, and as the toxic cloud moved through the densely populated areas around the plant, thousands of people were mortally poisoned. Tens of thousands more would suffer life-altering injuries (Varma & Varma, 2005). The exact casualty figures are unknown, and estimates vary depending on the source. Somewhere between 4,000 and 25,000 people were killed. Many of them died in the years after the event as a result of their exposure to the toxic gases that night (Eckerman, 2005).

---

*Fault finding versus fact finding*

The key goal of any investigation is to discover what happened so that we can prevent future similar situations. Because of this, I personally don't care whose "fault" any particular accident was. In fact, I don't even think of "fault" as a real thing. Investigations should always stick to just the factual information, with a keen eye turned toward links in the causation chain that can be broken. The more links you break, the less likely it is the same thing can happen again. Because of this philosophy, I really don't care which of the two causation theories is correct in regard to Bhopal. There are many similar root cause links in both causation chains that can, and have been, addressed on a system-wide scale through improved industry and regulatory standards and practices.

---

There are a huge number of resources, theories, histories, etc. online regarding Bhopal, many of them contradictory to each other. The two generally accepted causation hypotheses are that water was introduced into the MIC tank because of leaking valves and untrained operators, or that the event occurred due to sabotage (Eckerman, 2005). Rather than argue which of these might be correct, it is most productive for our purposes to focus on the known facts, and how they led to changes in chemical process safety.

The first important fact is that there had been multiple leaks and releases of toxic chemicals at the facility prior to the disastrous release. These leaks had occurred in various locations, involving various systems. In reading about the leaks, it is reasonable to draw the conclusion that plant employees were not adequately trained in emergency response, leak prevention, release detection, or use of personal protective equipment (Eckerman, 2005). I have serious doubts about whether they actually knew how to properly operate their equipment under normal circumstances, let alone how to react safely to a leak.

The second important fact is that there were people living in close proximity to the plant. India is no shining beacon of wonderful zoning laws, and frankly I am surprised that this type of disaster hasn't happened more frequently there given the living conditions of the poor. The proximity of a vulnerable population was exasperated by the fact that the plant did not have a useable, understood catastrophe plan or adequate public warning system (Varma & Varma, 2005).

The third important fact is that the safety systems in the plant were obviously inadequate. Some of the safety systems that could have reduced the effects of the release or possibly stopped it were shut off or otherwise not in use for various reasons. Investigations found that the plant did not receive enough investment capital to operate safely. Alarms didn't work, and the proper process for cleaning lines with water was not followed the night of the accident (Eckerman, 2005).

Fact number four is that safety audits completed in May 1982 identified thirty major hazards in the MIC unit. Investigators found that the items fixed after the 1982 audit had deteriorated again by the time of the 1984 release (Eckerman, 2005).

> *The importance of sabotage*
> As safety professionals, it is important to account for every causal factor possible when designing, auditing, or otherwise dealing with any system. If a system can be sabotaged, that contingency must be prevented just like any other risk factor. If you haven't learned enough about security risks to be confident in your competency, use outside resources and make security part of your continuing education plan.

In the end, there is little difference in the root cause findings of either causation hypothesis. The plant was obviously dangerous, the employees were obviously not trained, and the systems were obviously not built safely. The Bhopal tragedy is an example of a major management failure. Even if a saboteur did introduce water into the tank, a series of management failures was the root cause of the disaster.

As is oftentimes the case, corporate leaders decided that it was cheaper to produce a product in a poor area of the world, partly because they would not be held to first-world safety standards. As late as 2011, there were still high levels of pollution at the site (Tinsley & Ansell, 2011); I am unaware of any serious cleanup efforts that have taken place to clean the water and soil of Bhopal after 2011.

In the years after Bhopal, laws regarding chemical safety in the USA changed. The Emergency Planning and Community Right-to-Know Act (EPCRA) was passed in 1986 and signed by Ronald Reagan. Reagan was famously anti-regulation, and his signing of the bill indicates the popular support it had after Bhopal. EPCRA requires notification of the community about certain information to allow for more effective emergency response. Most importantly, the act requires notification of a release of hazardous chemicals. Safety professionals are well aware of the details of EPCRA, as we are the ones who do most of the paperwork.

> *Environmental regulations*
> Many employers understand that the "E" in EHS stands for "environmental." As such, most safety professionals have at least some environmental duties. It enhances your odds of gainful employment to maintain a level of competency regarding environmental standards.

It would be incomplete to discuss the long-term effect Bhopal has had on the safety profession without bringing up a major failure of the laws created in the United States in the wake of the tragedy. On April 17, 2013 a fertilizer plant in West, Texas, caught fire and exploded. Because emergency responders were dangerously close to the structure when it exploded, fifteen people were killed. The laws in place did not and could not prevent the fire and explosion; but more importantly, they did not ensure that responders knew what chemicals they were dealing with (Fernandez, 2014). The fertilizer that exploded is the same material used as the main component of the most commonly used legal explosive in the United States (ANFO). As a combat engineer, I used to use it on explosives ranges in training. It is an impressively powerful material.

There are some people who look at Bhopal and West and think the solution is more regulation. I see things a different way. We have plenty of regulation in place, it just needs to be used more effectively. I have worked in EHS for more than fifteen years, and I can't keep up with all the permits, reports, etc., etc., etc., that I am required to manage. The answer is shifting new resources away from passing ineffective laws and toward creating more tools to organize and simplify the requirements of the existing laws.

---

*Politicians and hot dogs are made of the same two ingredients*
It takes a lot of effort to change anything in the government. Set aside your political feelings for a few moments and focus on what will most effectively prevent another incident like the explosion in West, Texas. If we can make it easier and cheaper for companies to comply with the tangle of laws regarding EHS, the more likely it will be that those companies will comply. America can make it happen, we just need to make the investment.

---

For instance, the fertilizer plant in West reported to the EPA in 2012 that there were over half a million tons of ammonium nitrate stored at the facility, but the Department of Homeland Security was not notified (Fernandez & Schwartz, 2013). By streamlining systems and having a single unified website that provides a simple, straight forward way to report hazardous chemical inventories one time to all concerned agencies, populating and submitting all required permits, the ammonium nitrate report to DHS may not have been missed by local responders. This website should ping local emergency response agencies with information they need to know, and have a section dedicated to information that residents can look up regarding chemicals in their area.

The West and Bhopal cases present two opportunities for safety professionals in the future. First, they are a reminder that we have an obligation to be the counterbalance against a too-narrow focus by upper management on operating efficiencies that may hinder safety. Had the safety department at Union Carbide been able to effectively identify and mitigate the problems at the Bhopal plant, the incident may not have ever happened, or its negative effects may have been greatly reduced. Second, we can use technology to our advantage in simplifying regulatory requirements while increasing efficiencies and effectiveness. Creating a way to unify reporting requirements and simplifying the dissemination of information should not be difficult (just don't hire the web developer who messed up the medical insurance exchange website for crying out loud!). As developing nations mature to the point of creating safety and environmental law, this is an opportunity for safety professionals to expand our reach to a global market, consulting with foreign governments to create effective, modern systems that save lives, the environment, and property.

# 12 If you think safety is number one, I've got a bridge to sell ya

Unless you work at a nonprofit that focuses solely on making the world a safer place (i.e., ASSE, The National Safety Council, NFPA, etc.) safety is NOT, nor *should* it be, number one at your organization. For most safety professionals (myself included), *profit* is and should be number one. For others, providing service to our nation, community, schools, or other groups is and should be number one.

Employees know safety isn't number one, and every time we pretend that it is, we lose credibility. The "we" in this case is management and the EHS department. Employees know that if the company goes an entire year with zero serious injuries and makes a grand total of $3, management will go berserk, people will be fired, plants will be closed, and the CEO will jump out of an $85^{th}$ story window. Contrast that to a year where there were a handful of serious injuries (this can, sadly, even include a fatality or two) but the company had its first ever year of over a billion dollars in profit. Management smiles and pats each other on the back, every employee gets a Christmas gift, bagels, and a jacket, there are rumors of expanding operations, and the CEO receives a bonus that could pay the salaries of about twenty people who actually do all the work on the shop floor.

Then we put up a poster that says safety is number one and employees laugh at the safety program. Have you ever seen the movie *Carrie*, starring Sissy Spacek? Picture the scene where her mother says "They're all gonna laugh at you!" when Carrie says she is going to prom. That's what I am saying right now about putting up that "Safety is #1!" poster. Employees blow these posters off or make a joke about safety being number two. With the exception of a hopeful dreamer here and there, the employees know what the real number one is: Profit.

---

*Cultural references*
I refer to a lot of movies, TV shows, famous people, historical events, etc. I've always done this, because I found out early in my career that people will readily relate to cultural references quickly, and they tend (in my experience) to retain cultural references in their memory better because the reference and safety item create a neural link. It also makes for a good relationship builder, because now I have a common experience (watching the move *Carrie*, for instance) with the people I am speaking with.

Instead of wasting time and money on a bunch of posters and banners that lie to employees and ruin the credibility of your safety program, try this phrase sometime: *A strong dedication to safety is the number one way in which we will outperform our competitors to make the best possible profit.* It is an honest message that does not make employees feel like they are four-year-old children who are being told that Santa is only coming over this Christmas Eve if they are well behaved. This idea that we can blatantly lie to employees about the top priority of a company is belittling and patronizing to them.

It's time to quit lying not just to other people but to ourselves. You and I both know that if the company has zero incidents but only makes $3 this year, the EHS department is going to be one of the first places where cuts happen (and I don't mean the kind that require a bandage). Our job as safety professionals is not to make safety number one, it is to provide the opportunity for management to create an environment where safety is as important to profit as quality, customer service, and successful deadline management. For those of you who are ready to tear this book or your e-reader in half right now due to rage, take a deep breath. How can you or I, or *any* safety professional, make a difference in anyone's life if we have been laid off? What good is presenting the workforce with a mythical, obviously untrue statement? To win the game, we need to play the game. Be real, get real, and know the true meaning of what we do.

Open up annual reports that are sent to shareholders. Because the pool of stock ownership has widened, there are now efforts to include some EHS information in the reports so everyone feels good about their portfolios. Take a look at what's in there, I'll wait…

Lots of flavor of the month stuff, right? Probably a picture of a big, happy bear with a salmon in its mouth, or a tree photoshopped to be overly green. Anything in there about the number of facilities that have monthly safety committee meetings? How about a statement from the CEO about her commitment to exceeding OSHA standards? A paragraph explaining the various safety issues that were identified and solved in the past year, a breakdown of how many employees provided safety suggestions, a graph with the average response time to safety suggestions at each plant, or a brief sidebar about the commitment to using the hierarchy of controls? No? What do you think is number one?

Now take another deep breath, because this is going to get a little more real. You are not the most important person in the company. I'm sorry to be the one to tell you. It doesn't matter if you are a plant-level safety technician, the global EHS manager, or the CEO. The truth is that there is NO most important person in the company. It's a queenless hive, and every one of us is interchangeable. Don't believe me? Steve Jobs, generally regarded to be one of the world's most influential business leaders, died in 2011. How's Apple doing? Pretty good from what I hear. Jobs, Walt Disney, George Washington, and a host of other remarkable, planet-changing individuals have all passed on or retired from their organization and what has been the result? Apple, Disney, and the USA have all continued to thrive after their departure (this statement pending the next presidential election).

*Leading versus lagging indicators*

Another thing you probably saw in the annual report is the number of injuries, or an injury rate figure. Things that measure numbers or rates of incidents are called "lagging indicators" and are akin to looking back in history to measure performance. A serious weakness of lagging indicators is that they do not tell you *why* the numbers are what they are. Did the company get lucky, or did they do everything right? "Leading indicators" are a better measure of how well a safety program is performing, because they measure the things that a company does that lead to stronger safety programs. The items listed in the paragraphs above (the things that are probably missing from the annual report you looked at) are examples of leading indicators.

Why do I bring our lack of importance up? Because it is integral to understanding the concept that safety is not number one. A focus on a strong organization that is set up for long-term success in meeting its goals will allow the safety professional time and resources needed to help build or strengthen a culture. Steve Jobs couldn't have made Apple what it is today without Zen master Kobun, who helped him shape his philosophies on life (Brennan, 2013). Walt Disney had Ubbe Iwerks, who created *Oswald the Lucky Rabbit*, giving Disney his first big success and teaching him the value of an iconic character (Gablet, 2006). For George Washington, it was Alexander Hamilton, a genius at administrative and financial systems who wrote much of the Constitution, created the first national bank, and advised Washington on how the presidency should be held (Chernow, 2005). In all of these organizations an important figure worked behind the scenes to help create a culture of success that was sustainable.

Now, think of your company and our new idea that safety is the number one way in which we will make the best possible profit (or for those of you in non-profit organizations, the best possible way to most effectively serve our community). By focusing on the end goal of profit (or providing community service), you are speaking the language that leadership likes to hear. If the culture truly changes to make safety a high operational priority, the senior leadership team will begin to see the importance of safety and how important it is to profit. Boom, you've made long-term change. On the company scale, you may not be seen as the Jobs, Disney, or Washington; but in the EHS department you might.

Some time when you hear an employee say "Safety is number one, right?" try explaining this concept of profit being number one, and safety being the number one way that we make a strong profit. When I've done this, eyes grow large and people look astounded. A few have even told me, "Geeze, don't say that to management!" Part of the negative reputation that clings to safety nerds is that we are unrealistic. By admitting to employees that you are aware of the top priority in the company and qualifying that knowledge with the fact that safety gets us to our number one priority, you are buying credibility.

*C-Suite and SLT*
The C-suite is the suite of offices where the people whose job titles start with a "C" are housed. CEO, CFO, COO, etc. "SLT" is the Senior Leadership Team, which includes the C-suite and anyone else at the highest levels of management. In general, I try to fly below the radars of these groups. If you have to interact with either group, get some advice from a safety veteran or trusted coworker first.

Getting the C-suite to admit that safety is not number one is more of a challenge. The SLT is afraid that admitting where safety actually sits in the order of priorities will make them look cold, heartless, unfeeling... But they have to be that way. It's similar to when a safety professional investigates a horrible accident. To see the big picture clearly, the safety professional must be able to throw a switch that replaces the blood in their veins with ice, so that they can take a truly unbiased approach to finding incident causation. If the SLT gets too emotional in their decision making, they may not do what is best for the company as a whole. It may be hundreds, thousands, or even millions of stakeholders, from community members to employees to shareholders to customers who rely on the SLT to make emotion-free decisions based on the best logic and reasoning they can muster. This earns them a bit of forgiveness on the "safety is number one" issue, so don't be too hard on them. Well, they can fire you, so don't be too hard on them for that reason as well.

*Rainbows and puppy dogs*
Don't waste your safety budget on motivational posters. I worked in a facility where there were lovely posters of rainbows, puppies, hang gliders, etc. with safety messages printed in calligraphy underneath. Giant waste of money. Remember that hierarchy of controls we talked about in an earlier box? Posters are not even a little tiny part of that.

If the safety professional starts taking down all those posters talking about safety as the number one priority, there are going to be some employees who get upset. Taking down posters is change, and admitting that safety is not number one can make people feel like the company has just given up, and safety isn't a priority at all. Because of this, taking down the old posters should be followed by communications that educate employees on why safety is the best way to achieve the most profit. Talk about it in safety meetings, add it to your monthly safety bulletin, ask the CEO to send an email blast to the company distribution list and mention it at her next quarterly update. "We need to make money to stay in business, and we will make a lot more money if we work safely" should be a mantra that each employee can recite by the time the marketing campaign is finished.

Still think that safety is number one? Let's talk about that bridge...

# 13 Identifying the organization's many swords of Damocles

If you aren't familiar with Roman mythology, don't worry; I've updated the story of Damocles for you. Damocles was just hired on as the safety manager for a large construction company. He was an ambitious young guy, and wanted to someday transition to more of an operations management position. He shared this wish with the company owner, a tough boss named Dionysius Jr., who invited him to watch over the office one day while the owner was visiting the biggest construction site to date—A new theme park that was called "Gladiatorville." The site was huge, and in addition to all of the building materials needed, the construction company was fabricating and installing all of the gladiator-themed props and decorations. The morning that he was asked to sit in the owner's chair and manage the office, Damocles couldn't be happier. He had the pretty secretary, Juno, bring him coffee with sugar and cream. The maintenance manager, Quirinus, was happy to adjust the air in the office to match perfectly to Damocles' needs. Just when things were at their best, Damocles leaned back and looked up through the skylight directly above the owner's desk and saw a giant sword suspended directly over his head by a crane, held to the crane's hook by only a kite string. Damocles immediately jumped out of the chair, called the owner, and told him he never wanted to manage the office again.

---

*From safety manager to CEO*
It just doesn't happen. Safety is seen as a niche, and it is incredibly rare for someone to make a transition from the Safety Department to the C-suite. Personally, I think it is because we are seen as not understanding that profit is the number one goal of the company.

---

Okay, so you and I both know this wouldn't happen in real life. Safety people never become business managers! But the real moral of the story is that those in power also have a great deal of responsibility (even Spiderman knows that). This is true for everyone on your company's senior leadership team. Part of our role as safety professionals is to identify the swords of Damocles that are constantly lingering over the SLT's heads, and to replace the kite strings with heavy-duty chain. The swords never go away, they just get hung more securely.

For our purposes, the swords are employee fatalities, huge chemical releases, catastrophes that displace or kill community members, OSHA fines, and a myriad of other potential disasters. How can we, with our limited power and influence, help the SLT identify their swords of Damocles and then secure them in such a way that they are far less likely to fall upon anyone's head?

> *Just how expensive is this solution?*
> Everything has a value associated with it. By understanding the values involved, you can present a more concise, economically-driven argument for safety. Be honest with yourself; if you look at the numbers and your idea for a safety solution doesn't make economic sense, find another solution. Presenting something that is economically infeasible is usually a waste of your and management's time.

The first step is to combine your expertise with the tribal knowledge of the employees through Gemba walks. Once you talk to employees about what they do for a living, why they do it the way they do, and what they perceive as the risks involved, you can use your technical expertise to balance that input with a technically sound risk assessment. Risk assessments should be as objective as possible. Hard data is difficult to argue with. Which approach to the SLT sounds better to you? "Based on my observations, I think that our highest potential for a major loss is due to a car crash," or "Using the company's risk assessment tool, I have calculated that we are four times more likely to experience a fatality from a car crash than from the next most likely cause." Both arguments will likely hinge on more than this opening statement, but providing data based on a method of objective calculation is much stronger.

By securing the swords with something stronger than kite string, the safety professional is slowly building a stronger relationship of trust and mutual understanding with the SLT. Actions speak louder than words, and when our actions illustrate that we understand some of the challenges the SLT face each day, the SLT should begin to be more receptive to safety in return. This give and take approach involves compromise and understanding on both sides, but just like any relationship one party has to get the ball rolling.

The lesson of the Sword of Damocles is really about empathy for your SLT. While we safety professionals are focused on the moral obligation of a business to provide a workplace free of recognized hazards that does not pollute the environment, the SLT is focused on a much wider range of moral obligations. A few of the SLT's obligations outside of safety are:

- to provide a profit to shareholders;
- to provide continued stable employment to employees;
- to provide economic stability and growth to communities;
- to maintain a positive image of the company;
- to provide a product or service to the community/nation/world.

> *SLT pocketbooks*
> The first time you talk to your SLT using economics as part of your decision-making discussion, they may tell you that they will worry about the money. That's great, and it shows that they will likely invest heavily to make sure safety remains a top priority. It is still important for you to keep economic considerations in play as you do your background research; every SLT has their dollar limit.

Capitalism, economic growth, competition... these are all good things. Effecting change, not just in an organization, but in the world, must involve incentives that are focused on capitalistic principles. When the safety professional approaches the SLT (or any other manager/management group for that matter), it is important to have a well-rounded argument for safety that includes more than just regulatory requirements or moral obligations. Competitive edge must be part of the argument, not just because money talks but because there is a moral perspective to money talking.

We all like to say that money is the root of all evil, but money is also the thing you get paid with. Your paychecks, annual bonus, electric bill, mortgage, children's education, and the beer in your fridge are all part of the monetary system that the oligarchy of American CEOs have a moral obligation to keep healthy and robust. As safety professionals we are an important cog in this economic machine; money is not a bad thing, and companies striving to earn more of it is not a bad thing (just like it isn't a bad thing for a safety professional to seek out a higher salary).

The bad thing is greed. Here's where the blurry line lies. Because the line is blurry, and oftentimes curved and flexing like a snake in response to the pressures exerted upon it, safety professionals must always approach issues with a "big picture" perspective. Safety nerds need to balance our concern for getting employees proper workstations against the sword dangling over the senior leadership team's head that might be labeled something like "budgeted expenses." If we show empathy and understanding to the senior leadership team when the resources simply are not available today for something that can wait until tomorrow, upper management will be more likely to respond to us with empathy and understanding when there is no room for compromise with an issue because it is an immediate risk to life and limb.

There are real challenges to this. Many of the SLT are of an income status that disconnects them from the real world. You and I, humble safety professionals, are worried about how we can save enough money for our kids to go to college without saving so much that they don't get any financial aid. SLT members can oftentimes sell enough stock options at this very minute to pay for an education at the school of their child's choice. We cut back on our satellite package and beer budget so that we can make a car payment; they have enough left over from the sale of those stock options to buy a new car with cash so that their kid can use their old Lexus at school. Money means different things to people who have more of it at any given point than you and I together will have over our entire lifetimes.

> *The magic formula*
> Sales required = Accident cost/Profit margin.
> Make sure you use the profit margin as a percentage; a 6 percent profit margin gets plugged into the formula as .06.

Most safety professionals have been taught this formula: The value of units sold to pay for an incident equals the cost of the incident divided by the profit margin. Does your company sell industrial fan blades at a seven percent margin? Cool. If an employee amputates a finger for a total cost of $12,000 in medical and lost time (forget about the hidden costs for now), that means that your company has to sell $171,428.58 in fan blades to pay for the incident.

> *Hidden costs*
> Most EHS professionals have been trained that hidden costs are everything besides medical and lost time costs. They are also called "indirect costs." The idea is that hidden costs can't be measured, and there are a lot of very artistic iceberg models that indicate hidden costs compared to the visible costs are much larger and can sink any ship that Kate Winslet sails upon. Mmmmm, Kate Winslet... Oh, sorry. Got a little distracted. The concept of hidden costs goes all the way back to our friend Heinrich, and hidden costs have been shown to be about four times the direct cost of an incident.

Now, how about those hidden costs? Most are actually measurable given the time and data. Here's a few of them:

- Downtime from the incident. You can measure how long the equipment involved was down. Your company has calculated the cost of downtime of every line and process; when they engineered the manufacturing process, they figured out how much idle time costs.
- Cleanup costs. Time taken to clean up and decontaminate the blood, chunks of bone, and all those other goodies safety nerds like to call PIM (potentially infectious materials), multiplied by the hourly cost of the labor. If there is no PIM, it can be cleaning up crashed boxes, destroyed equipment, etc. If you're really good you can even figure out the cost of each cleanup component and the cost of disposal of any hazardous waste.
- Scrap. This is an easy one. How many fan blades need to get scrapped because of the incident? The same engineers who figured out downtime costs can tell you how much this material costs. Sometimes, the material can get looped back into production and sometimes it can't; either way there is an associated expense.
- Overtime. Someone's got to cover for the employee who is now out of work for a while after the incident. Easy to calculate this if you can get the hourly labor rate.

- Production dip. All of those employees who saw Sam get his finger chopped off are going to work a little more slowly for a while; they're going to spend time talking about the incident, which is going to result in longer breaks (I call this the "Give a Mouse a Cookie" effect).

You can probably come up with many more "hidden" costs that can be calculated or estimated to a great degree of accuracy. Now calculate all this (or simply use the 4:1 ratio) and present the findings to upper management. How many more fan blades will you have to sell to pay for the incident?

When a person inevitably says that a few accidents are the cost of doing business, I like to respond by saying that it is one of the more controllable costs. "The cost of doing business" is the label on one of those swords. Shrink that sword, while illustrating that safety improvements carry over into other positive improvements, like higher production rates, less downtime, greater efficiencies, etc. and you will have a better chance at getting approval for projects.

# 14 The genus python of the safety world

When I was fourteen I, through a series of circumstances that could be its own chapter in another book, ended up alone for a few hours in the house of some people I didn't know in far-away Kansas City the day before Christmas Eve, rather than in my rural Minnesota home with my family (I'd have to write the whole story as fiction because nobody would believe it). The home owners had a large glass tank with a ten-foot python in it. I looked at it, and when it looked back at me I realized just how small and tender (like veal) a boy of my age was. After the residents (honest to God, the guy's name was Monty) returned home and we were eating frozen bananas together (seriously), we had a few questions for each other. I was surprised when I learned that the word "python" refers to a whole family of snakes, not just a single species. I'm not much of a snake expert myself, so I did a little research on pythons. What I found out is that there are twelve recognized species of python. I remembered learning when I was very young that pythons don't have venomous fangs. I wondered if they had teeth, but thankfully the python that night didn't offer to show me his open mouth.

It turns out that pythons do have teeth; while not full of venom, their teeth can actually be quite nasty. Pythons use their teeth to grab their prey, then they wrap their bodies around it and squeeze until the prey is dead. When the prey is dead, the python swallows it whole (Roy, Campbell, & Toure McDiarmid, 1999). I've seen some pictures of python teeth, and they look like they can really do some damage. Granted, they aren't like tiger's teeth that are engineered to tear a roast-sized chunk out of another animal, but they are sharp and scary.

OSHA reminds me of a python. They've got teeth, but not teeth that are likely to kill you. Rather, OSHA uses its teeth to hold you so the squeezing can start. The teeth are the monetary fines that come with citations. The squeezing is the application of future actions, like being put on the Severe Violators Enforcement Program (SVEP). If you've been involved with a SVEP action, you remember the way that OSHA's squeeze felt.

Most citations involve a fine that is a pittance, even after their recent increase. Financially, the fines don't mean much to most employers; they are easily reduced, and they don't come close to the cost of an incident that could occur given the identified hazard. For example, an unguarded sprocket and chain might garner a serious violation for a grand total of $12,471. Which is an employer naturally going to have a greater response to: The $12,471 fine, or the potential amputation that will cost tens of thousands of dollars?

---

*OSHA inspection priorities and citations*
From OSHA (n.d.a.)
OSHA comes to visit employers based on the following priorities:

1. Imminent danger situations: OSHA witnesses a hazard that could cause an immediate death or serious injury. OSHA inspectors might be on their way somewhere else, witness the hazard, and pull over to perform an inspection.
2. Fatalities and catastrophes: If someone dies, or three or more people are hospitalized, OSHA will come and say hello.
3. Complaints: If an employee contacts OSHA to make a complaint about their workplace not following OSHA rules, an inspection will result.
4. Referral: In my experience, most referrals are from media reports about an incident, although sometimes another government agency will contact OSHA to report an issue they witnessed while on the premises.
5. Follow-ups: Checking back on a previous citation to make sure the employer is abating the hazard as ordered.
6. Planned or programmed investigations: OSHA, after all of the above priorities, still has time to randomly come calling on employers. They try to stick with employers in industries that tend to be more hazardous.

Citations and penalties, from most to least serious:

1. Willful: The employer knew about the hazard and did not take appropriate action in plain indifference to the safety of employees.
2. Repeat: The employer had been previously cited for the same thing and is now being cited for it again (for multi-establishment employers, the two citations do not have to be at the same establishment).
3. Serious: A violation was found that could have harmed an employee.
4. Other-than-serious: A violation of OSHA standards was found, but it is not a serious threat to employee well-being.

---

Okay, wait… before you answer that, remember that we are talking about the C-suite here. I and most other safety professionals have seen upper management hem and haw on simple safety changes that could prevent a hugely expensive injury, then go nuts when OSHA writes them a ticket for less than the cost of fabricating an expanded metal cover. They understand how the python effect works.

We're back to those python teeth. To the untrained eye, they look nasty and frightening. To a herpetologist (someone who studies *snakes*, not conditions that are acquired during trips to Las Vegas), the teeth are nowhere near as dangerous as most laypeople might think when they see them. Just like to the untrained eye, OSHA citations might look worse than they really are. The dangerous thing is the squeeze that comes if you don't get the teeth out fast enough.

You can look up citations on the OSHA website, which is where the following information is from. OSHA fined a Dollar General store $122,100 in May 2015 resulting from a November 2014 inspection. OSHA claims the store, located in Bear, Delaware, had boxes and merchandise stacked in front of exits, along with other violations. Over one hundred thousand dollars in fines might seem steep to a layperson, until they realize that this citation package was part of the OSHA python squeeze. According to OSHA, Dollar General had a six-year history of similar violations at stores around the country. Dollar General did not correct the hazards on a system-wide scale (i.e., pull out the python's teeth), and the squeeze followed. Some of the earlier violations included a locked exit in a Wolcott, New York store after discovering the same hazard in a Buffalo, New York store in 2010. How can a business have such a systemic problem?

I do not and have not ever worked for Dollar General, and I can't speculate on their internal risk management practices. Nor is Dollar General the only dollar-themed retailer to have repeated problems complying with OSHA standards. I find it hard to imagine how or why various dollar stores have not taken the steps needed to address the repeated problems at their stores, and am curious to know why the cited violation keeps occurring. For all I know, Dollar General has done everything in their power to address the issue, but somehow can't quite get their arms around it enough to curtail it. I've worked as a corporate safety manager for another retail company, but thankfully we did not face this sort of issue.

I have, however, worked with a company that did not understand how repeat violations worked. The concept of repeat violations being possible across state lines, at businesses with local names rather than the corporate name, was not an easy one for upper management to fully comprehend. I lectured them on it multiple times after OSHA visits, wanting to pull out the python's teeth and avoid the squeeze. It took some convincing, but they began to understand. Unfortunately, local management across the country did not quite get the message in time.

The result of my experience was that the company began to get targeted more often after a few repeat violations were issued. It didn't take long before upper management really felt the squeeze of this, and implemented needed, sweeping changes. It wasn't my statements of the possible consequences as much as it was the scary python that motivated management change.

---

*Contesting an OSHA citation*
Employers can contest any OSHA citations. The process usually starts with an informal conference, and can evolve from there. If your employer is going to contest a citation, make sure corporate counsel (and probably a good consultant) are on the team. Most of the time, contesting a citation results in reduced penalties.

Another part of the OSHA python squeeze is the press release. Negative headlines really get to management. Here are a couple of examples of how press releases on the OSHA website are titled: "Dollar Tree Store endangers workers again in Texas," "A. Hyatt Ball Co. Inc. exposes employees to fire, explosion, other hazards," and "Health Care Products Inc. exposes production workers to amputation hazards." These headlines are rarely if ever followed by stories that talk about how the citations were reduced or erased, or how the company in question successfully abated hazards and became a safety leader in their industry (on the limited occasion where that happens).

It is as if the leadership at OSHA is a python with fang envy, and I can't say I blame the agency. *Variety* reported that a 2012 newscast that accidentally included a sexually explicit image in the background of a camera shot (it was a small popup window on a computer screen) resulted in a fine of $325,000. Dollar General, according to the OSHA citation from November 2014, exposed five employees to blocked exits, accounting for $60,500 of their total citation amount. The worst case scenario is that a fire occurs and all five employees are killed (along with a number of customers). By this reasoning, an employer would have to place 27 employees in mortal danger to reach the same fine as the television station. But that isn't how it works, because even if an employer *did* place 27 employees in mortal danger, they would still likely only receive a $124,709 fine (the limit for repeat or willful violations) as a result. So what the government is telling us is that little Johnny seeing private parts used in naughty ways on TV is a worse crime than placing employees' lives at risk of death.

As of this writing, OSHA's penalty amounts have been increased to reflect inflation. This is only the second time OSHA's citation amounts have been adjusted. According to testimony provided to the Subcommittee on Workforce Protections by Dr. David Michaels during the "Protecting America's Workers Act" debates in congress prior to this adjustment (Dr. Michaels is the OSHA administrator at the time of this writing), not keeping up with inflation meant that "…the real dollar value of OSHA penalties [has been reduced] by close to 40 percent." Those teeth are looking a lot smaller now, aren't they? The C-suites of America don't get upset with OSHA because of the fines; they get upset with OSHA because it means *they got caught*.

As the safety profession continues to strive for more accountability and greater professionalism, it is important to remember that companies are not as worried about citation amounts as they are about getting caught. When groups like the Chamber of Commerce seek to pull out some of the OSHA python's teeth, it weakens our profession as a whole, and it illustrates just where these groups feel safety belongs. Changes in business ethics are occurring right now because companies see that investors want their money to be placed with organizations that operate with ethics as a guiding force. To fully create an environment that is ready for real advancement in safety, investors have to look beyond the obvious and examine the groups that businesses choose to be a part of, and judge a company's ethics based on the stances those associations take on ethical issues such as employee safety.

# 15  That funny little scar on my lip

The M113 armored personnel carrier (APC) began service in 1962. It is shaped like a big camouflage breadbox, with light aluminum armor so that it can ford waterways and is more efficiently transported by air. The APC rides on two tracks, is equipped with a .50 caliber machine gun, and can carry over a dozen soldiers into and back out of battle. It has a top speed of 37 miles per hour, and can stop on a dime… if it hits a big enough tree (Wikipedia, n.d.b.).

I discovered this remarkable stopping ability while on a training course that ran through the woods of Camp Ripley, Minnesota (also known as "Poison Ivy Central"). I had driven an APC dozens of times, and had a ton of hours of experience on the beastly green aluminum box. I was still in college, and hadn't yet learned the safety value of regular training drives that were observed by an instructor, so I was a little irritated that I had to go through the exercise. At the halfway point, I would stop and get in the back. Another soldier, a former high school quarterback named Todd, would switch places with me and drive the training course back to our origination point. One of my buddies, a guy named Jake whose girlfriend I had accidentally asked on a date once, was in the APC behind me. I wanted to own some bragging rights, so when it was time to go through the heavily wooded training course I took off like a shot, hoping Jake would take my unspoken dare and race through the course with me. Later, Todd told me he was wondering what in the hell was going on with my speed, and Jake said that when I took off so fast he knew what I was up to and simply shook his head, thinking I was nuts.

I left Jake's APC in the dust and the instructor, who rode in the track commander turret above and behind my position in the driver's compartment, nervously asked over the mic if I was sure this was a safe speed. I assured him I knew what I was doing.

I raced through the woods, and reached an area where the left side of the trail bordered a steep slope. I wanted to avoid the slope; the APC is prone to rollovers, which often ends up killing the track commander and driver, who are propped out of the APC. The trainer (a lieutenant fresh out of officer candidate school) told me to guide to the left because I was too close to the trees on the right, but I ignored him. I wasn't going to slow down, I knew what I was doing, and that slope on my left was spooking me.

The trainer was just beginning to get irritated with me, telling me again to guide left, when we went from about 25 miles per hour to zero, all at once. As Jake tells the story, "I saw Scott take off like a bat out of hell. About five minutes into the trail, I saw the tallest tree in the woods start to sway back and forth, and I knew he hit it."

Todd, riding in the troop compartment of the APC, was not belted in. The seats in back face inward, so that the left or right sides of the troops in the back face the front of the vehicle. When the APC stopped, Todd kept moving at 25 miles per hour, hitting the engine compartment hard and bouncing to the floor. He was banged up pretty good. The trainer in the command turret hit the turret hard enough to knock the wind out of him and bruise his ribs. As karma would have it, I got it the worst.

The driver's compartment is built so the turret opening is at about mouth level. The lap belt will keep a driver from getting thrown out of the vehicle in most rollovers, but doesn't do a thing to prevent contact with the turret in the event of a head-on crash. I slammed into the turret mouth-first. The mic on my helmet was forced past my teeth (luckily it only loosened a couple and didn't knock any out) and ended up in the back of my mouth. My bottom lip was pretty much smashed like a worm run over on a driveway, and I felt thick blood running down my chin onto my uniform where it soaked in.

As the lieutenant behind me sucked wind, I reached up with a shaking hand and pulled the bloody mic out of my mouth and spit out blood. After a few moments, we spoke to each other to ensure we were okay; he checked on Todd after that. Within two or three minutes of the crash, I was driving again, noting that the gigantic pine tree I hit had a big chunk knocked out of it. Jake caught up to me soon after as I drove very slowly through the rest of the trail.

When it was time to switch drivers, Todd was only mad at me for a few moments. Once he saw my mangled mouth, his justifiable anger with me faded. I apologized and he said he hoped I'd be alright. When I got into the back of the APC, I laid on my side on a bench and let the blood from my mouth run onto the floor. My mouth seared with throbbing pain.

When we returned, a bunch of the guys from my unit were waiting around; they had either gone through the course already or were waiting to go through on an upcoming run. When the training runs were over, we would pile into the two APCs and head back to our tents in another part of the camp. They saw me stumble out of the back of my APC and had a million questions. Todd, Jake, and I recounted the story from our various points of view while I found a mirror and checked out my torn-up lip.

I am not and was not at the time, a smoker. However, one of my buddies convinced me that after a rough experience like a crash, a cigarette might help relax me a bit. What he nor I thought about was how fast my body would absorb nicotine when the cigarette was up against a ripped open lip. I took about two drags and fell over, nearly throwing up and too dizzy to stand. It was not relaxing.

I've crashed a lot of stuff in my life, but this was the worst. I learned a couple of lessons from it that I use to this day. I always tell drivers about it during training. I find it helps them accept the road test as a necessary evil. And I never underestimate the damage that can be done at speeds under 30 miles per hour. The APC was completely undamaged; it is built to take a harsher beating than I could dole out, but the bodies of the soldiers involved took a bashing. I am never surprised when learning about serious injuries sustained in a low speed crash. I also use the story to remind myself of how people in big machines can turn into cave dwellers quickly and commit major errors. A forklift isn't too different from an APC, and it pays to train and observe over and over again, because that much mass at a low speed can easily cause a bad incident.

# 16 The four personality types of safety committee members and how to engage them

Vince Lombardi should be the hero of every safety committee chairperson. I'll need to have a few drinks of scotch tonight, because I am a lifelong Vikings fan about to expound on the greatness of a Packers coach. Uff-da. Here goes.

> ### The importance of sports
> In my experience, sports references are even more important than cultural references. People tend to at least know a bit about a few sports, and being able to relate a safety topic to sports can help bring about impressive retention levels when you are conveying information to employees or managers. I once had a manager who needed to educate upper management about everyone in our department, so he made "baseball cards" of all of us, complete with stats and facts. It was a big hit (har dee har har).

Lombardi led the Packers to five NFL championships in seven years. He's generally considered one of, if not THE, greatest coach in the history of the game. The trophy awarded to the Superbowl champion every year is named after him. When Lombardi came to the Packers as head coach in 1959, the team was coming off a one win, ten loss, and one tie record. The next season, Lombardi's first as head coach, the team finished with seven wins and five losses (Maraniss, 1999). How could Lombardi lead the team to such a turnaround?

Vince Lombardi was first and foremost a fair man and a hard worker who stressed the importance of effort instead of a focus on failures (leading metrics, anyone?). In an age where racism was common in sports, he did not tolerate racism among his team or the fans. He had enough clout in the town of Green Bay that he did not tolerate businesses who discriminated, and forced cultural change outside of the stadium. After going on to coach the Redskins, Lombardi showed the same level of fairness and respect to gay players, at one point telling an assistant coach "I want you to get on [underperforming gay team member Ray] McDonald and work on him and work on him—and if I hear *one* of you people make a reference to his manhood, you'll be out of here before your ass hits the ground." He was also fair in his expectations of players. Players were given a set of goals, and while Lombardi held his players to high standards, he held himself to equally high (and in many cases higher) standards. In this manner, he became an agent of change. When those who are led see their leader as honest, fair, and not self-exulting, it becomes far easier to bring out the best in the team as a whole (Maraniss, 1999).

Vince had a blocking scheme that you can use with your safety committees. It is called "run to daylight" and it means that blockers on the offensive line are assigned an area to block instead of an individual defensive player. When the running back gets the ball, he is instructed to run to daylight; in other words, look at where the blockers have made a clear path for you to run through, and then run the ball through said path (Maraniss, 1999). You will have some members of the committee who have the energy to pick up the ball and run to daylight if they are given a little prodding and direction. That motivated person is your running back, and the other members are the linemen, "opening a hole" by providing support, fresh ideas, and motivation. Once this begins, you will be amazed at how many safety "touchdowns" you can get out of employees who are not safety professionals.

Safety committees are a far cry from a professional football team. I oftentimes look at the various state laws and standards regarding safety committee membership and have to shake my head. For example, my home state of Minnesota includes language that "employee representatives [to the safety committee] shall be selected by their peers." And according to Washington state OSHA rules, members to the safety committee must be employee-elected. Yeah, because in Minnesota and Washington there are so many people clamoring to get on the safety committee that we have to hold elections. Right.

---

*Unrealistic rules*
Unfortunately, unrealistic rules are a part of the safety world. Follow them as best you can, and when you have to improvise, make sure you are providing as high or a higher level of protection than that intended by the rule.

---

Here's how "elections" to the safety committee actually happen: The safety professional informs the management team that a safety committee is being established or there is a vacancy, and that members should represent each department. The safety professional then says something to the effect of "Employees are supposed to choose their committee representatives, so see if anyone wants to do it; if there are multiple interested parties, have the department vote." The managers go back to their departments and ask for volunteers, and things get interesting. The second safety committee personality type, the "Activists" jump all over it (you, the Safety Professional, are the first personality type). If there are no Activists, then the third personality type, the "Good Citizens" *might* volunteer. If there are no Activists or Good Citizens, the manager will just pick the next type of safety committee member, better known as the "Piranha Cattle."

In the end, most safety committees have a mix of personalities, and if you are flexible enough to work with this diverse group your safety committee can be successful, meaningful, and yes, even a little fun.

Personality type one is you, the Safety Professional. You have a full plate, and the safety committee is one more thing to plan and prep for. It can be tough to get excited for the meetings, because you already have enough going on to keep you busy, and the meetings usually devolve into complaint sessions. It is an hour a month that you would rather be doing something that actually makes a difference (like getting your teeth drilled).

OR…

You love these meetings. The committee members are involved, they have great ideas on how to make the workplace safer, and it is an hour where you get to actually spend time talking about safety solutions and getting things done. Oh, and there's cookies!

---

*Motivation for attendees*

Whether it is a safety committee meeting, annual training, or any other meeting you have to lead in safety, figure out what will make it less tedious and boring to people. Think about a department that holds mandatory meetings you hate: Okay, for everyone else that is you. How can you get people to show up and listen? Find a way and you've taken a big step toward more acceptance of the safety program.

---

The second personality type on the safety committee is the Activist. Insert your heavy sigh here, because at your facility, you probably know these folks by name already. The safety committee is, to them, one of two things: It is either their way of airing every grievance they've ever had about the company, or it is their approach to finally get That One Thing changed. You know That One Thing; it's the situation that the Activist thinks is the biggest safety issue on earth. It might be number 7,846,371 on your priority list (or not related to safety in any way to anyone except the Activist), it is a huge pain to work on because there is no time or budget to address it, and the worst incident that could happen is that a cricket might stub his toe; but to the Activist, That One Thing is THE ISSUE.

Personality type three is the Good Citizen. These are usually your best members of the safety committee, and if you have a lot of these folks, you probably love your meetings. They bring good ideas to the table, rather than just showing up to complain or silently put in their time, and they genuinely care about getting results. They've joined the safety committee because they feel it is the right thing to do. Safety goes beyond just following the rules for the Good Citizens and for many of them, there is a personal experience (like a past injury to themselves or a loved one) that has taught them the true value of safety. Beware, however of the Garrulous Good Citizen; they will take over the meetings and make them feel more like a poorly planned Toastmasters event.

*The bad egg*
Sometimes a Piranha Cattle can be a bad egg, someone the department manager wants to get out of their hair for an hour a month. If you get a bad egg that is ruining your meetings, they need to be told that they can participate in a positive manner; they can sit down, be quiet and do their time; or they can leave.

Personality type four is the Piranha Cattle. These are the poor souls who have been "voluntold" to be on the committee. They might have even had an election in their department to see who *must* go to the committee. They usually don't have a high participation level, and show up primarily to stay out of trouble with Da Boss. Sure, the cookies and pop are nice bait to keep them showing up, and they might enjoy some time in a cushy meeting room instead of busting their knuckles on the floor, but remember that these people have been *drafted* into service. If you really want to get the most out of them, it's going to take some creativity (we'll get into that in a minute).

How does the safety professional take this dynamic cluster of personalities and turn them into a functioning team? Believe it or not, it is possible. There are roughly eighty zillion guides to a successful safety committee out there in the world that you can buy or find online for free, and most of them are actually pretty helpful. If you want to look at a few and find one that seems to fit your situation, go for it. Just keep this key concept in mind: You are building a team.

The safety committee is more like a junior high football team than a pro squad. You (the coach) can't really pick and choose who your players are; they come to you when they sign the sheet saying they want to play. You might end up with eleven offensive tackles and no quarterback. Your job is to learn the personality of each member, and use what you learn to mold the team into a cohesive unit (even if that means your wide receiver is built like a defensive lineman and your punter is afraid of the ball). As Vince Lombardi said, "The measure of who we are is what we do with what we have. (Maraniss, 1999).

Start with yourself. You've just become the nucleus of the team... the coach... the platoon sergeant. If they aren't standing with you and believing in what you are trying to accomplish, your ship is sunk before it ever sets sail. So you have to believe in yourself first, and in the possibility that this team can achieve its goals—then make an accomplishment of the team's goal the *expectation* for everyone. If the safety committee meeting is not your favorite part of the job, ask yourself why, and address that issue the same way you address the root cause of an incident. Find or create a positive reason to attend the meetings, and begin looking forward to them. It might be a break from the normal grind, a special form of sugar incentive you use to lure in the members each month, or even an hour to talk to an interesting employee that you usually don't get to spend time with. Set yourself up for success by making the meeting time that you will enjoy.

After you have yourself excited for the meetings, start to meet on an individual basis with the members that the managers have sent you. Make this an informal "getting to know you" activity and stop by their workstation (this is a good time for a Gemba walk, isn't it?). You'll discover who your different personality types are pretty quickly, and you can begin thinking about how to guide the meetings so that they aren't taken over by a Loquacious Larry or dragged down into a complaint-fest by a Negative Nellie.

It can be a trick to keep everyone involved, so create various forms of incentive for positive action on the part of committee members. Start with your Good Citizens. They are likely to fill key roles in a positive manner, and can be encouraged to take leadership positions. Be careful not to load them down with too much extra work; remember that this is an additional (usually thankless) activity on top of their already full plate. Let them know how much their involvement is appreciated. One other note on the Good Citizens—they can sometimes be an Activist in disguise, so until you know them well, don't rest too much faith on them.

Now that you have some leadership support in place, take a look at your Activists. Start to work on short-, medium-, and long-term success gates toward their issues of concern (unless the issue can be solved quickly and easily, in which case just get it done or outline the reasons it has to wait). This is a great time to huddle with management or maintenance team members on these issues; they will appreciate that you are getting the Activist off their back, and will therefore oftentimes be willing to work with you on a strategy for addressing the issue.

---

*The importance of being Vern*
Sometimes, after an Activist has aired their grievance, they will settle down. Please remember that their important issue still exists, and needs to be addressed by you in the same manner as it would be if they were bugging you about it every week. Being a good listener cuts down on the number of personality conflicts you become part of. Bonus points if you get the reference in the subject line above.

---

If you have spent any amount of time with an Activist before, you know that it is imperative that you validate their feelings about the issue and let them know that you are working toward a solution. If they can see a few small steps happening that move the ultimate goal closer to fruition, they will oftentimes start to feel like they have begun making a difference. It will take all of your reasoning and diplomacy skills, but you need to convince them that the timeline you have in place is appropriate. Be empathetic and understanding. Remember that although the Activists can be a pain in the butt, they are human beings just like you and I; the things that are important to us may not be important to others, but we still want others to respect how we feel. The Activists are the same, they just have different priorities and personal approaches. The positive side to Activists is that if you are able to get them to believe in you as a safety leader, they can become Good Citizens and can be extremely useful when it comes time for other employee involvement activities.

When Teddy Roosevelt and his son Kermit explored the Amazon in 1913, some local villagers trapped a school of piranha and starved them for days prior to Roosevelt's visit. When the former American president arrived, the locals told him the legend of piranha cattle, or sacrificing one cow to keep the deadly fish busy while the rest of the herd crossed the river. The locals then forced a cow into the water. Roosevelt was impressed by the way the vicious piranha ate the cow down to its skeleton in minutes (it was like watching the action at a buffet in a Packers-themed sports bar) (Millard, 2009).

Your last personality type on the safety committee is the Piranha Cattle. The rest of their department has thrown them into the river first to keep the safety piranha busy so everyone else can all get their job done without any bites. Seriously... This is how many people think of safety committee meetings. I know we safety professionals feel like the most important people in the company; after all we make sure everyone goes home with the same number of fingers and toes they came to work with. But we have to be realistic here and admit to ourselves that no matter how much we would love it if the rest of our company laid rose petals at our feet, safety is just never going to be a glamorous job. Your Piranha Cattle members are going to be the toughest crowd to engage and keep positive, and in a lot of cases, they will make up the largest number of safety committee members.

Engaging the Piranha Cattle starts with getting to know them personally, just like everyone else. Where you needed diplomacy and empathy for the Activists, you'll need to be a slick (but honest) politician for the Piranha Cattle. If they talk about their kids a lot, maybe you could add a discussion topic to the meetings about safety at home. If they show up to work late every day and try to sneak out early, they may be more responsive to a meeting style that includes time for members to socialize and talk about anything *other* than work for a few minutes before getting down to business. And finally, never underestimate sugar motivation. Hold the meetings over the lunch hour and order everyone food, or hold the meetings after lunch and provide cookies and pop (as well as healthy snacks for your health-conscious members). The trick is to make the Piranha Cattle feel like there is something in the meeting that they like, instead of feeling the sharp teeth of the safety piranha gnawing at their leg. The more comfortable they feel, the more likely it is they will become positively engaged.

These techniques will not work for everyone. You will always have a Good Citizen who is determined to make long speeches, an Activist who cannot be pleased no matter what you do, and Piranha Cattle who spend the entire meeting playing Angry Birds or checking their Facebook page. The only element you can control is your own attitude, so be sure to make the meetings engaging for yourself.

Marv (name changed to protect the verbose) was a Good Citizen who liked to talk. A lot. If you asked him what time it was, he'd tell you how to build a watch. He would preface every question with an opening statement, so long I thought he was rehearsing for testimony before congress. The guy was killing my safety committee by talking it to death. Other committee members began hating him, and after meetings I would overhear people complaining about

Marv. Marv would wear his glasses low on his nose and look over them, giving others the impression that he was looking down at them. When he would start a speech, there was even one safety committee member (a Good Citizen, no less) who would actually stuff ear plugs in his ears. Marv irritated everyone with his speeches, including me. It took him so long to get to his point, that I oftentimes was baffled, and had no clue what the actual point was supposed to be. I was afraid to ask for clarification, because we'd be back at square one. The confusion on my part was what led me to the solution for Marv. One meeting when he started talking, I stopped him early. "Marv," I said, "I can't always follow you real well, so I am going to bullet out your points as we go." I had my laptop projecting as the meeting went so that everyone could see the notes as I took them. Marv started talking and I typed out a bullet, then stopped him. "Does that look about right, Marv?"

He nodded. "Yeah, I think so."

Before he could get started again, I said, "Okay, everyone. This is a good discussion point." I turned to the guy who had just popped in one earplug and asked, "What do you think about this?" We ended up starting a discussion that kept Marv in the background. He was satisfied, because he had made a meaningful contribution to the group, and the group was happy because they were able to put in a word here and there.

I've had a lot of Piranha Cattle over the years, but Jim (name changed to protect the indifferent) was one of my favorites. Not only did he make it completely obvious to everyone that he was bored and hated being at the meetings, but he was tired of having pizza for lunch each month *and* was always trying to steer the conversation toward the ways management had screwed the union in the last contract. I had tried getting to know Jim but he didn't like me. I knew he was a hunter, so I added a few funny photos to my slides that were wildlife or hunting themed. He hated them and told me to quit being so damned goofy. I had finally given up on Jim, until the end of a meeting one day. Everyone had filed out and gone back to work, but Jim hung around. I was confused and quite honestly, a bit apprehensive. He walked up to me, and I prepared to be called a jackass and lectured. Again.

"About those racks," he said. We had an issue in the plant with rolling racks that were used to move partially assembled products from station to station as they were assembled. One of the results was that the end station had a huge pile up of racks that blocked an emergency exit walkway. We had been working on the issue in the safety committee for months.

"Yeah," I said, a bit hesitantly. I was ready for him to tell me it was time to fix the flipping problem or just shut up.

Instead, he said, "I think that if we moved Jack's machine to the other side of the walkway, we might be able to fit them into the corner. We could probably route them along the back wall there to get around the other traffic on their way back to receiving."

"Can you show me?" I asked. He did, and he was right. It was a brilliantly simple idea.

At the next meeting, I wanted everyone to know that Jim had figured out the solution to our problem. I said, "Hey everyone, we need to hand it to Jim. He's the one who came up with the idea to clear the walkway that had been full of racks."

Everyone congratulated Jim. Jim looked at me and scowled. After the meeting he called me a jackass. It finally hit me (I can be a little slow sometimes) that Jim just doesn't like attention. A couple of days after the meeting, I found him at his workstation, and I told him that his idea to move the machine was brilliant. "I also understand now that you don't like being the center of attention," I added. "Sorry I called you out."

Jim shrugged and mumbled something I couldn't understand. I doubt it was a compliment.

"Hey, if you have any more ideas like that, you can just drop by my office or hit me up after a meeting. I'll always be happy to take credit for your brilliant thinking."

Jim smiled in spite of himself and said, "Yeah, you would." He still seemed tuned out during meetings, but when he sat there pretending to read the newspaper, he was actually doing a lot of thinking. He came up with a few more gems, but like many Piranha Cattle, when his year on the committee was up, he did not volunteer to come back.

Sheila (name changed to protect me from this person tracking me down to tell me how upset she is that I used her real name) hated working overtime. What she hated even more was working forced overtime on the night shift. She was an Activist, and her distaste for forced overtime had little or nothing to do with safety. However, Sheila was creative. She mulled the overtime issue over and over in her mind until she found a way to try to make it about safety. If she had to work overtime, she was extra tired when she drove home after work and that was unsafe. It was especially bad after working overtime on the night shift. She volunteered for the safety committee at a time when there were no open seats. In an effort to encourage future employee involvement, I allowed her to join after clearing it with her manager.

---

*When an employee calls OSHA*

Employees have every right to call OSHA and remain anonymous. Most of the time, it doesn't take rocket science to figure out who it was; in my experience, it is oftentimes someone who was recently let go. If the person still works for the company, it is of the highest importance that no punitive actions are taken, even if the employee completely made up a hazard to get OSHA to come. Retaliatory actions will get you and your company into serious trouble! If an employee threatens to call OSHA, my standard response is "You have every right to do that. OSHA's contact information is on the bulletin board with all of our regulatory posters."

---

Sheila spent her entire first meeting talking about forced overtime. Another member would bring up a burned out light by a staircase, and Sheila talked about overtime. Someone else said that the MSDS binder in the shop was getting ratty, and Sheila talked about overtime. I asked what people wanted for lunch next month, and Sheila brought up overtime. Each time it came up, I said, "You know, Sheila, that sounds like it might be more of an issue to bring to your union steward. There really isn't anything I can do about it." She was mad at me, and insisted that if she fell asleep while driving home, it would be my fault personally for not making the forced overtime stop. *And* she would be sure to tell everyone that I had blown off her concerns. She even threatened to call OSHA.

I told her I understood that working overtime was a pain, and tried to empathize by mentioning that I had just put in a few extra hours the previous Friday. She didn't care, and kept insisting that I make prevention of overtime my top priority.

Now might be a good time to mention that Sheila and I worked at a prison. I had safety issues beyond the extreme, because not only was I protecting employees, prisoners, and visitors, but because we had an industrial shop on the premises, a high school, and most of the facilities were over a century old. Overtime was not an issue I could afford to make a priority.

The next month, Sheila came to the meeting and complained about overtime again. How could it be legal to *force* people to work overtime, she wondered. I wanted to tell her that the prisoners probably weren't going to guard themselves, and the state was enforcing a hiring freeze. Yes, more guards and less overtime would be nice, but it just wasn't going to happen.

---

*Unions*

It is important to stay in communication with a trusted HR contact. Use them as a sounding board to make sure issues brought to you aren't union grievance issues masked behind safety. I have been in several situations where a union tries to unfairly place me in the middle of a labor dispute.

---

After the meeting, I told Sheila that I was going to bring the overtime issue up to the Captain of the Guards and to the union steward again, but it had to stop coming up in the safety committee meetings. It was not a comfortable conversation, but I approached it as diplomatically as I could. She did not understand at all, and stormed off. I secretly hoped she wouldn't show up to another safety committee meeting.

Next month, there she was. She said nothing during the whole meeting, and acted more like a Piranha Cattle than an Activist. After the meeting, I asked her if the Captain had talked to her.

"Yeah," she said. "There's nothing I can do about the overtime. I give up. You know," she added with a sigh, "I just want to get home with enough energy to get my kids ready for school or to make them dinner."

Her activism instantly made more sense to me. I told her that I didn't have any kids yet, but I hoped that when I did someday I'd be able to help them out like she did when I got home from work. She smiled and thanked me, then walked off. Over the course of the following months, she became more of a Good Citizen in the meetings; I'd like to think it was because I finally connected with her regarding her family issue.

OK, Coach. You've got your players. You've got your challenges. Now go out there, form a team, and score a few safety touchdowns. Don't be afraid to fail. After all, as Vince Lombardi put it, "The price of success is hard work, dedication to the job at hand, and the determination that whether we win or lose, we have applied the best of ourselves to the task at hand" (Maraniss, 1999).

# 17  Brett

The first fatality I investigated occurred at about 9 am on a pleasant Tuesday in June. I was in the corporate office, working on something that seemed important until a coworker answered our department phone and told me that a grain elevator in Kansas was on the line. They had a worker go missing that morning, they had found him in a bin of corn a few minutes ago, and he was unresponsive.

---

*Incident command*
During an incident, it is important to have a plan in place. FEMA.gov has training that will help prepare you. Ensure that your managers and reports get some training too, so they do not take actions contradictory to the incident plan.

---

Switching immediately into emergency management mode and remembering FEMA's National Incident Management System guidelines, I took over incident command on our end of the situation. The local safety leader's voice quivered as he told me what he knew. Brett, the employee, had been missing since about 8. The other four elevator employees went looking for him. One of them noticed that a grate was missing in the head house atop the bins. The searching employee shined a flashlight down the hole and sixty feet below, where the pile of corn peaked, there was a lifeless body.

---

*When tragedy strikes*
Make sure you have a support system of experts you can rely on. Your employer's insurance company is a good start. The loss control staff there has a lot of experience, and can help you navigate through overwhelming situations. It isn't a bad idea to talk with them proactively about how to handle a major incident so you are ready when the day comes.

---

The safety leader continued to call with updates as they happened, and by the time I was at the airport waiting to get on my flight to Denver, Brett's death was confirmed. I had my notes in front of me, and I studied them over and over during the flight. It was another three hour drive to the small town with a hotel closest to the tiny hamlet where the incident had occurred. By the time I checked in, it was close to midnight.

The next morning, I was at the regional manager's office before 7. The safety leader, regional manager, and I talked about the situation for a few minutes before driving to the elevator together. An OSHA investigator arrived a few minutes before we did, so my first view of the actual scene began with an OSHA walkthrough. We answered all the OSHA investigator's questions, took all the pictures the OSHA investigator took, and explained what we knew.

---

*Dads at risk*

I once found statistics related to how much more likely a new father is to be hurt at work, but could not find them again while writing this book. As a dad, I know that for the first couple of months after having a baby, I was completely out of sorts due to stress, lack of sleep, and a focus on keeping my family's needs met. Give new dads a break, and know they may be at higher risk.

---

While the OSHA investigator interviewed employees, I spent time at the accident scene on my own. My interviews would have to wait. I looked at everything time and time again, taking pictures of every inch of the head house (the building that sits atop a row of grain silos). By the time I was able to interview employees, I had a hypothesis.

---

*Interviewing witnesses*

There is an art to interviewing witnesses. Be gentle, take lots of notes, and remember that the people you are interviewing have gone through a horrible event and are scared, tired, and in shock. Their friend is dead or badly hurt; draw information out of them with a soft touch, asking lots of open ended questions and being the best listener you can be. Take breaks as you need to, to avoid getting emotionally overwhelmed; this process will be tough on you.

---

In the interviews, I spent a lot of time asking the employees to tell me about Brett, and a little time asking them about the incident. I learned about his three-week-old daughter, and how he would come in on Saturdays sometimes just to make sure everything was locked up properly. I also learned enough about the grates and operations to firm up my conclusions as to what happened. I needed to make sure there was no evidence in the corn, and spent hours sifting through about fourteen thousand bushels. I found a dead mouse and a rock, but nothing that added or took away from my ideas of what probably happened.

It was after 10 that evening when I returned to my hotel room and typed up an initial report, which I emailed to my boss and several company executives at about one in the morning. They all responded before I went to sleep that night at 2.

If you do this job long enough, you will lose an employee. During those few days I spent in Kansas, I was only partly a safety professional. The rest of the time I was a counselor, listening to a group of people tell me about their deceased friend. There is no real preparation for a situation like an employee fatality; if you have an experienced peer, talk to them during the event for advice and a friendly ear.

> *It is never anyone's "time to go"*
> Anytime I hear someone imply that the Hand of Providence was the cause of a workplace death, I lose my mind a little. Every workplace fatality is a tragedy that probably could have been prevented. Passing off a tragedy to an unseen force is passing the buck. Find the root cause as if your life depends on it, because somebody else's does.

The process of investigating a death at work provides huge learning opportunities. This particular incident came back to a root cause of poor design. When all was said and done, we spent tens—maybe hundreds—of thousands of dollars identifying every floor opening at hundreds of grain elevators in over a dozen states, and bolting the grated covers to the floor using the basic principles of machine guard fastening. I also learned that every day the systems and designs I helped create sent a lot of people home safe; even so, one failure out of a few thousand chances was not good enough.

The technical lessons of the investigation were experiential. I did all the things I had been taught to do, like taking a lot of pictures, controlling the scene tightly, and recording everything; but it was the "soft" lessons that meant the most to me. Human beings stand apart from other animals due to emotion, and I never would have gotten useful interview information had I not connected to each person I talked with on an emotional level. It meant a lot to them to be able to tell me about their relationship with Brett. Who he was mattered to his friends at work just as much as finding out why he died. The safety leader for the small cooperative knew Brett well. I patted the large man's shoulder and asked how he was doing. His normally ruddy face was white, and he stared out of the window at something a million miles away. He nodded and said, "It's tough."

The Kansas wind eventually blew the tough days away, and for the next couple of years, I called the safety leader on the anniversary of Brett's death. "How you doing?" I'd ask.

The answer was always, "It's a tough day, Scott."

# 18  $12 million says a lot

After the investigation of Brett's death wrapped up and we had finished dealing with OSHA, the project to secure floor openings gained traction quickly. Nothing grabs upper management's attention like an employee fatality. This was an opportunity to strike while the iron was hot, and although it may sound cold blooded, I wanted to use Brett's death to my advantage to make more safety improvements quickly, while there was a lot of motivation for change.

> *Circular floor openings*
> A circular opening creates a situation where the cover cannot be turned and dropped through the opening. Whenever possible, design floor openings to be circular, like a manhole. Secure the guard over the opening so that it requires a special tool for removal, and then educate the hell out of employees about why they need to respect the opening.

I had seen that throughout the various grain handling facilities in the company, there were a great number of fall-from-height hazards. I brought this to the attention of upper management. It wasn't the first time I had done so, but in the aftermath of the fatality they were ready to listen. I wanted to do something radical: Send out a ton of manpower to identify every single walking and working surface that was not in compliance, then bring that walking or working surface *into* compliance. I was clear that we were at high risk for another fatality due to a fall from height, and not only would we lose another employee, but because the risk had now been identified by multiple sources, we had no acceptable defense after the next fatality. My request was accepted.

> *Fixed ladders*
> Fixed ladders are ladders that are permanently attached to a structure.

We used the division's safety staff as well as the division's auditing staff and scoured the hundreds of locations around the country for walking and working surface hazards, using a checklist I had built especially for the task. Every stairway, every fixed ladder, every railing, every catwalk, and every floor was examined over the course of several months. Repairs were ordered as deficiencies were found, and middle and low level managers were informed that capital requests related to safety—and this project in particular—would not be denied.

---

*Millwrights*
A millwright is a person who builds and maintains machinery. They are typically adept at welding. When you run into a large project like the one described in this chapter, they are your best friend.

---

To understand the true scope and breadth of this project, it is important to remember that we are talking about 300–400 grain elevators that were built between 1920 and the 2000s; some by contractors dedicated to doing the job right and some by local farmers who jerry-rigged half the building together. We kept millwrights busy for weeks at a time, and in some cases entirely redesigned portions of facilities. It was, in a word, epic.

---

*Manlifts*
There are three primary types of manlifts in a grain elevator:

1. Electric lifts. These are similar to a small elevator.
2. Rope manlifts. These are counter-weighted wood platforms with a guard rail and a foot break. Pull the rope one way to go up, the other way to go down. Try not to let go.
3. Endless loop belt lifts. These are the things of nightmares. They are a belt that never stops moving, and has a foot platform and hand hold every so often. You step on and off as they move.

---

The employees watched with a certain level of curiosity. The company had always talked about safety as a priority, but nothing like this had ever taken place before. As catwalks, ladders, manlifts, and other infrastructure was improved, the employees and managers found that their jobs became easier. It was faster to get to places that needed attention around the facility, and employees no longer tried to get out of as much work, because it was now less dangerous and easier to do previously dreaded tasks. Efficiency improved, morale improved, and the facilities looked more business-like. Soon, as I would drive around the country visiting locations, I would see yellow dots on the fixed ladder runs from safety gates that had been installed at the rest platforms. Areas of facilities that were previously unattainable could now be reached for inspections, and people told me that they weren't as afraid to do their jobs as they had been in the past.

---

*Permit-required confined space*
There is a technical definition for this, but I'll spare you. A permit-required confined space is a confined space with hazards in it that make rescue dangerous and/or difficult. If you need the technical definition, please visit the OSHA website.

---

As the walking and working surface changes showed systemic improvements, we were given the opportunity to improve other areas as well. Boot pits, generally considered a permit-required confined space, were changed so they had a stairway rather than a ladder for entry and exit, and many facilities added ventilation to the pits. We discovered wiring deficiencies in areas that were classified due to combustible dust, and improvements were made there as well.

The more we made investments in safety, the more employees talked about safety. The more they talked about safety, the more likely they were to report injuries in a timely manner, make safety suggestions, show up to training, and wear their personal protective equipment. Regional business unit managers began making safety a higher priority, secure in the knowledge that capital requests for safety improvements would not be denied. When capital was requested for non-safety issues, managers looked closely, trying to find ways to tie the proposed expenditure to safety. In doing so, they started putting significant time into hazard identification and prevention through design, without ever having taken a class on it or getting the terms drilled into them.

> *Prevention through Design (PtD)*
> I love PtD! Work with engineers when you are designing facilities, processes, etc. to design safety into the system from the blueprint forward, and your life as a safety professional becomes much easier, because you've eliminated hazards before they ever were born. To learn more about PtD, visit the bookstore at www.asse.org and order yourself some reading material.

Success breeds success, and the project that began with a goal to mitigate floor opening risks ended up changing the way we did business. The walking and working surface improvements cost the company over $12 million. Nobody will ever know how many lives were saved, or if *any* were for that matter. But the company was innumerably improved.

As we made improvements, employees would tell me stories of the times they practically died due to a certain walking and working surface deficiency. There was the auditor who slid down the slippery sloped roof of a steel grain bin, but was blocked from falling by reaching up to grab a railing at the edge (we made toe boards standard, realizing the risk of falling under a railing, and started putting in more tie-off points on bin tops), the employee who bumped a railing only to have it fall off the catwalk he was on, the department manager who wouldn't use a certain fixed ladder because it was loose and wiggled, or any one of dozens of other stories. The stories were always followed by a statement about how happy the person was that we were making changes.

I learned something interesting during the walking and working surface risk mitigation project. The CEO and leadership team can talk safety all day, they can hold meetings and grade managers on their safety performance; but spending $12 million on safety improvements says and does more than any of those things to improve the culture. Employees know that money is number one. By *investing monetarily* in safety in such a dramatic and visible way, in every facility, management *illustrated* the importance of safety, and it changed the way employees worked quickly and effectively.

Culture change has to include some type of visible, major investment because the proof is in the pudding. I have a relative who has gone through the twelve steps of Alcoholics Anonymous. Apparently, part of the program is apologizing to people for the horrible things you've done to them while in an alcohol-induced haze. This relative made his apologies to me, and my response was that the words were meaningless. The things he'd done were unforgettable and in some cases unforgivable; but I would give him back some of the trust, respect, and familial care that he had lost if his actions reflected his words. He's done well in the many years since that conversation. It still bugs me when I run into someone from my home town who finds out that I am related to this person and tells me a story of how he ate a beer glass at the bar one night, but I no longer worry about panicked phone calls in the middle of the night or reading his name in the police blotter.

Changing a safety culture is similar. Words and corporate directives are nice, but actions, investments, and change that makes the workplace tangibly different for employees makes all the difference in the world.

# 19  Does this make me flotsam, or jetsam?

Things were shaky, at best. I was the corporate safety manager for the retail division of a Fortune 500 company that owned grocery stores and distributed groceries to independent retailers. Why were things shaky? Because retail grocery is a business cursed with inherent shakiness, running on about a 2 percent margin while a certain chain of pole shed-like supercenters out of Arkansas is driving competitors into the ground.

---

*Claims manager*
A claims manager is an employee who manages all of the insurance claims for the company. This includes workers' compensation, liability, auto, property, etc. Make friends with a claims manager, they can teach you a lot.

---

*Occupational health nurse*
Occupational health nurses work with claims managers to make sure injured employees are getting the treatment they need in an effective, efficient manner. They can also help figure out good light duty job options.

---

We had a great team. I loved the job, and it was one of the few times in my career I have really liked my boss a lot. His name was Bill, and he had worked in safety and risk management for decades. On the distribution side of the business were two corporate safety managers who were about the same age as me. My cohorts from distribution were talented and energetic, and our offices were huddled in the same hallway as the corporate claims manager and occupational health nurse. Informal collaboration was common, starting with one safety manager rolling his chair into the hall and asking a question to the other two. Soon, the three of us would have all our chairs in the hallway, talking through an issue that we were facing. I'd see Bill peek out of his office at the end of the hall and smile, pleased with the team he had assembled. With the occupational health nurse and the claims manager officed between us and Bill, questions about work comp, return-to-work issues, and liability exposures all ended up in hallway conversations. Ideas flourished, and it was the first time I was in the middle of true team synergism. Then the CEO found us.

*Getting your butt chewed*

It's gonna happen. Sorry. A common saying in safety is that if you aren't pissing anyone off, you probably aren't doing your job. If you get berated in a public space, turn your emotions off. You can cry, punch your file cabinet, swear, or whatever else you need to do later in private. It sucks, I've been there. When you're over the worst part of it, talk to a peer and engage in your favorite decompression activity.

*Competition*

You are going to become friends with safety peers from competitor companies; it just happens. Sharing information related to what has been successful in your company's safety program is usually OK; just check with your boss before you give away any information that might relate to competitive business practices.

In the interest of not being sued, I won't use the CEO's real name. Instead, let's call him "Dick." Dick was awful, both from a business sense and as a human being. He was the kind of CEO that gives all other CEO's a bad name. Dick would publicly and harshly chew people out in the middle of Cubeville; the rest of us would pretend to work as we listened and felt bad for the poor soul who was berated in the middle of a public space, crossing our fingers it wouldn't be one of us next. The coffee stations in the building all had the company's store brand of coffee, but since Dick didn't like the company brand, he had a special coffee station installed in his office with high-end coffee. If anyone else was caught drinking a brand of coffee other than our store brand, they were disciplined. We were housed in a five-story building, with each story having one set of restrooms. Dick decided he wanted a wet bar in his office on the third floor, so the restrooms on his floor were removed to install a wet bar and private restroom. All the other employees on the third floor now had to go to another floor of the building to use the restroom. On the business side, the company was barely able to stay afloat, there were rumors of questionable accounting practices, and we were getting killed by that store I mentioned earlier that is basically a giant pole shed. Our retail stores were dropping left and right, and the independently-owned stores that made up our distribution customers were trying to find ways to survive once the big box competition opened a store within twenty miles of them. Since leaving the company, Dick has gone on to slash and burn his way through other companies, leaving a trail of unemployment and long-term investor disappointment.

As is oftentimes the case when the CEO is a "turnaround specialist," the EHS department was identified as easily trimmed fat. It started when the claims manager, a tall, kind man in his early sixties, resigned and took a position on the east coast. We were all disappointed and asked him why he was leaving. First he told us that he wanted to be closer to his elderly parents so he could help take care of them, but as time crept closer to his last day with us, he also confided that "Things are going to change, guys. The CEO is going to make big cuts. Keep your resumes updated, and be ready for anything. It's going to get ugly around here." It chilled us all, and we wondered what he knew and how he knew it (I never did find out those details).

A few weeks later, Bill was not in his office one morning. There was no indication on his calendar that he was out, and he hadn't said anything to any of us about using a vacation day. It wasn't like him. Early in the afternoon, Bill's boss Kathy, the corporate legal counsel, arrived in our hallway and asked if she could speak with us for a few minutes. We met in one of the offices, and Kathy informed us that she had fired Bill that morning. We were now reporting directly to her. We asked why Bill had been let go, but she only told us that "it was a tough decision." I called Bill when I left work that night and talked to him for a bit. He told me the firing had been political, but he didn't think he should go into details. The one piece of advice he offered was "Get out of there, Scotty. The place is bad news."

I couldn't believe it. Even with the shaky profit margin, this was the best job I had ever had to that point. I loved the people I worked with, I loved going out to the retail stores, and I had been racking up the safety wins for months. I wished Bill the best, and promised to stay in touch (which I did for years before his passing). After hanging up, I convinced myself that even though it was the complete opposite of everything I knew about Bill's personality, his warning must have just been sour grapes. That was wishful thinking, and it wasn't long before I discovered I was wrong.

The corporate counsel was an OK supervisor. Kathy only remembered the EHS department existed when it was time to deliver bad news, and for the next few weeks we were free to do our jobs with very little interference from above. Flying below the radar didn't last long, however. One day she was back in our hallway wanting to talk with us all, a feeling of dread draped around her like the Grim Reaper's cloak. She informed us that we were being moved out of our offices and into cubes near the accounting department. After she left, we all agreed that things smelled bad. I finally started to face the music, and began updating my resume. As we transitioned from our quiet, spacious offices to our loud, cramped cubes, the company began selling off retail stores. An email was sent out informing all of us at the corporate office that even though retail was being "differently sized," there were no plans for workforce reductions at the corporate office. Right.

---

*Layoffs and work comp claims*
Once rumors of a layoff start, work comp claims often take an upswing, as people begin to worry that their medical issues will not be covered if they get laid off. The same rules apply in this situation as in other questionable claim situations. Let your insurance company handle the questionable claims.

---

It was our last day in our offices, a Friday, and we were complaining about the change. All of us wondered who would be the one that forgot about the location change and walked into his empty, silent office on Monday morning out of habit. One of the other safety managers said he was really disappointed that he couldn't use the mini-fridge in the small break room that we had been sharing with the corporate security managers. "Nobody else ever uses it," he said. "It's a shame. Now I'll have to find a new place to put my lunch bag every day."

I decided to stay late that night and do some work to help the department's transition to cubes feel a little less uncomfortable. When the building was mostly empty, I went to the break room, scooped up the mini-fridge, and lugged it into the elevator and down to the second floor cubes. The lights were low and I knew the building was practically empty, so I hauled the fridge to my cube, where I discreetly tucked it under my desk and plugged it in. On Monday morning, I showed the other two safety managers, and we decided to keep our fridge secret to avoid any trouble or overcrowding. People walking by couldn't see it unless they ducked into my cube. We had lost our offices, but we had gained our own private mini-fridge. By Tuesday afternoon, it was full of snacks, water, and pop.

Three days later, as I was working on my computer in the small cube I now called home, I heard a booming voice behind me say, "Well, there it is!"

I turned and saw the corporate security director standing inside my cube with his hands on his hips, staring at the mini-fridge.

I replied with the smartest thing I could think of, "Ummmm…"

He chuckled. "I bet you didn't know that refrigerator belonged to me. I bought it and brought it in a few years ago, then my wife quit making me sandwiches for work so I never used it anymore."

"Oh, gosh. I didn't realize it was yours. I thought it belonged to the company."

He shook his head. "Don't worry about it. They really did you boys wrong. If my fridge makes it a little easier, then by all means, you guys take it. I'm really sorry about everything you've had to go through the past couple of months."

"Thanks."

"Tell you what, Buddy. It ain't gonna start getting any better, but it's sure gonna get worse. Take care of yourself." With that, he was off again.

Two Mondays later, I took the day off to sleep in and do some projects around the house. When I came in Tuesday morning, I picked up my phone and dialed into the voicemail system to see if I had missed any messages. The system didn't recognize my password. I tried a couple more times, thinking that maybe I just had fat, sleepy fingers, but the system would not recognize me. When I logged onto my computer I saw that the previous morning Kathy had sent me a meeting invite at 9:15 to meet at 9:30. She had never done anything like that before, and I started getting nervous.

When the other two safety managers came in, they looked relieved and told me they were happy I "had made it."

I asked what they meant, and they looked at each other, then back at me. "You haven't heard?"

"Heard what?"

"They laid off about three dozen people from Retail yesterday morning."

I slapped my palm to my forehead. Locked out of voicemail. Late notice on a meeting with the boss who only talks to me when there's bad news. I knew my job was over.

"Holy buckets. I'm getting laid off today." I explained the evidence I had, and although they assured me that the company couldn't possibly lay me off, I could see in their eyes that they didn't believe what they were saying.

I called my wife (the first of two; we'll refer to her as "Practice Wife" from this point on).

"Hey, Honey. I'm pretty sure that I'm getting laid off today."

"You better not! We can't afford that!" Practice Wife was not an empathetic soul.

"Well, there isn't much I can do about it."

"Oh, you just tell them 'no.' That you refuse to be laid off."

"Um, I'm pretty sure that isn't how it works."

She then spent about ten minutes telling me that I had better not come home without a job. The day I got laid off, I decided I had also had enough of Practice Wife and wanted to be single again soon. It was about to be a summer of change.

I sent an email reply to Kathy to say I had taken the previous day off. I had a vendor coming in at 9 to talk about a pest control contract, but I could meet her when that meeting was over. She replied with another meeting invite, for 10.

---

*Theresa from human resources*

Theresa didn't actually work for the company. I spent a lot of time in the HR Department. If you've seen the George Clooney movie *Up in the Air*, you will recognize that Theresa was a contractor who travels around the country laying people off. These contractors are very specialized and serve an important purpose in making sure companies follow all applicable laws when reducing the size of their workforce. You may not like their role, but they are filling a niche and making a living, just like you and me. At least they didn't become a telemarketer.

---

When the time came, I walked into her office. There was a woman I didn't recognize standing next to her. Kathy smiled politely and said "Good morning Scott. This is Theresa, from the Human Resources Department."

I sighed heavily and my shoulders dropped.

Kathy nodded her head and said, "Yeah. I'm sorry. It was ultimately my call of who to let go when the word came down, and it was a tough decision." It sounded eerily familiar to what she told us when she fired Bill.

The people I met with that day to file the paperwork that comes along with getting laid off told me I took it better than anyone else they had ever seen. I told them that was mostly because this wasn't actually much of a surprise. I had seen the writing on the wall and had been waiting for it to happen.

> *Bridge over the river layoff*
> Layoffs come with a lot of emotions and can be difficult to handle. Make sure you remember that people you work with today may be helpful to your chances of gaining future employment. Heck, you might even get invited back to the company that has foolishly laid you off once they see the error of their ways. Take it like an adult, show appreciation for the time they've given you, and wish everyone the best.

I was home by 11:30, and after making myself a tuna sandwich and bowl of cream of celery soup for lunch, I opened up my binder of business cards and starting calling every safety professional I had ever met, all the vendors I worked with who might have connections, and the headhunters I had made contact with over the years (this was before LinkedIn). By 3, a few had even called me back.

> *Networking*
> Networking is one of the most important tasks for a safety professional. Your network can help you with expertise, advice, job searches, program examples, and a plethora of other important tasks or knowledge items. I encourage all new safety professionals to join the American Society of Safety Engineers (we talked about them earlier). Go to www.asse.org for membership information.

My total time of unemployment was under two months, primarily because I used my networking resources to full advantage and kept a positive attitude. I made follow-up phone calls to people who I talked with at ASSE meetings, and people who had forgotten they ever knew me. I passed my resume out at ASSE meetings, and made arrangements to attend educational seminars some of the risk brokers in town put on monthly in order to network with their clients in case any were hiring. The lessons I learned were to be outgoing and friendly, to be unafraid to put my name out there, to network my hind end off at every opportunity, and to be open to selling myself to previously unknown potential employers. The week I accepted an offer for a loss control position with a third party administrator, I turned down a regional safety manager position with a transportation company. I started the new job excited to be in a new company, free and clear of Practice Wife, and working with a team of loss control experts. The layoff, in the end, was an extremely difficult experience that had a positive outcome simply because when I needed to find work, I sought it out aggressively and with the right attitude.

Today there are even more options for a safety professional who finds him- or herself out of work. The expansion of social networking tools like LinkedIn allows us to expand our circle of potential employers, headhunters, etc. exponentially. Advances in technology mean that there are new and novel methods of manufacturing that require equally new and novel approaches to safety. Had the layoff occurred today, I could have spent my time off learning more about mobile robotics, 3D printing, autonomous manufacturing, creating safety apps, or a myriad of other things to make myself more employable. These new manufacturing possibilities mean that more small shops are springing up, creating a need for more loss control staff at insurance companies and a niche for consultants to fill. The future is now, and when it comes to being ready for the future of the safety profession, Malcolm X said it best: "Education is the passport to the future, for tomorrow belongs to those who prepare for it today."

# 20 Safety discipline needs a spanking

> *Soft skills*
> Just as important as understanding OSHA standards, technical information, or engineering concepts are "soft skills." Soft skills are the things needed to relate to another human being on an emotional level.

Nothing kills a safety program faster than a "Safety Cop." I've worked in multiple companies where employees have been conditioned by past safety professionals to think that I am going to be a robotic disciplinarian who has no care for the actual conditions, needs, and pressure of their jobs. This is a problem for the safety profession, and it might be best to look at it using a root cause methodology, the "Five Why's" technique.

1   *Why* do people think I will be void of any personality and make them work in a manner they don't want, regardless of their reasons for bending or breaking safety rules?
    Answer: Because previous experience has taught them that this is how safety professionals, police, teachers, angry parents, etc. have behaved in the past regarding safety. When it comes to safety, it seems that most people have had more exposure to a history of quick punishment than a history of understanding or rewards for good behavior.
2   *Why* have people been punished when it comes to safety more than they have been understood or rewarded?
    Answer: The answer to this became much clearer to me once I had a child. I noticed that when my son was in or about to be in a situation where he could get hurt, I was likely to be aggressive with my words and actions in my attempt to keep him from harm. I wasn't going to follow him around, constantly saying "Good job not climbing up the back of the couch," or "Thanks for not approaching that strange dog," that would just give him ideas. My natural parental behavior is to react to a dangerous situation by issuing a command, and following up with a positive reward (e.g., if you do what I tell you, we can read a Clifford story) if the command is followed. Threats of a time out have almost no effect on the squirrely little monkey.

What my wife and I have learned about our son is that if we calmly react to the situation and then talk to him in an even-toned manner about why he should not do something, focusing on the positive alternative behavior and offering a reward for the desired behavior, he is much more likely to stop the undesired behavior. (Alert! Parental bias ahead!) Since my little all-star is perfect in every way, what else would I expect?

3    *Why* are people not rewarded more for good behaviors?
     Answer: Has a police officer ever pulled you over to tell you that you did a great job coming to a complete stop at that last red light? No. And if she did, you'd think she was really odd and would be upset that you had to delay your trip (and possibly speed to make up time after the stop). On top of that, it is impractical and unnatural to follow someone around, constantly rewarding them, and it is a lot of work to find something to reward someone for every time you see them. The natural result is to look for dangerous behaviors and then react to them.

4    So what the hell are safety professionals supposed to do to promote good behaviors and reduce bad ones? Yeah, I know; not a "why" question. So give me a time out.

---

*Welcome to the machine*
Here's why BBS should be lower on the hierarchy of controls than any other items. If I design a machine correctly and make 1000 copies of that machine, then present each copy with the same situation, I will get one result 1000 times. If I take 1000 people and present them with the same situation, I will get 1000 results one time each. We can control machines, but humans are a wild card.

---

Answer: Psychological research has shown that rewards work better than punishment (Belsky, 2008), but that loops us back to the conundrum of being stopped by a weird cop who wants to tell you how happy she is you are driving safely. The problem is that we are looking at rewards incorrectly. As a result, we've got a lot of behavior-based safety programs that are wasting your time and money. BBS is ineffective and a poor way to focus efforts in safety. When an employee sees someone walking up to them with a checklist in their hand, ready to observe their work, they change how they do things. It's called the Hawthorne Effect (Henslin, 2008). The better approach is using your newly-found Gemba walking skills to stand or sit next to an employee, with no checklist in your hand, and have a *conversation* with them about their job (and oftentimes their life in general). Believe it or not, there is a reward here. Human beings are social animals, and we really like it when a friendly person shows interest in what we are doing, talks to us about fishing, garden projects, and our kids; and smiles a lot the whole time. Not smiling in a creepy "You must be nice to me or I'll eat you" manner, just being a normal, friendly person.

5    *Why* does the answer in #4 sound like it isn't really work?
Answer: Remember when we talked about ergonomics? The most basic solution to ergonomic calculus is to mimic a natural environment. By approaching employees in a genuine, friendly, and positive manner you are doing what comes naturally to the human animal. In the workplace, we are so far removed from our natural environment that anything that takes our mind closer to a state of natural human behavior is felt as a reward.

Okay, we've touched on the basics. Punishment doesn't work that well, and rewards should be a genuine conversation rather than an artificial, possibly demeaning, pat on the back. Why do I say the pat on the back can be demeaning? Because too often, people feel like they have to come up with something to "reward" an employee for. So you get the out-of-touch safety manager who walks up to a guy in the tool room and says "Nice job wearing your steel toe shoes." Seriously? How would that safety manager feel if the plant manager came up to him and said "Nice job wearing a polo shirt in this strictly business casual environment." Might as well pat the tool room employee gently on the head, and say "Good boy, Squirt," and give him a lollipop.

How about this as a reward instead: The safety professional approaches the tool room employee, and immediately recognizes that the tool room employee knows about ten thousand times more information about machine shops than the safety professional does. So the safety professional says "Hey, Ken. That part you're making looks pretty cool. Is it complicated to machine something like that?"

Whoa, whoa, whoa, hold on! This conversation has nothing to do with safety! But it does. Ken is going to tell the safety professional a little bit about what he is doing and how he does it. If Ken is guarded at first, that's okay. The safety professional can take the small bit of information that Ken has given him, file it away mentally, then move on to "How did you get interested in tooling" or "Did you draw up the CAD design for that?" or, if Ken has loaded the part into a CNC and is now at his desk, "Hey, that's a nice bass!" while pointing out a picture of Ken with a huge fish.

---

*Open-ended questions*
Remember that teacher who caught you writing a note in class, and asked you "You know better than that, don't you?" She demanded an answer before she'd let it go, and you hated her for it. Don't be that person. Employees have reasons for doing what they do. Treat them with respect and ask questions that pull more information out of them so you can find the root cause of their behavior, while allowing them to save face.

---

At this point you might be thinking, "Okay, smarty pants, what about the time I walk up to Ken and he isn't wearing his safety glasses, or even worse, he's changing tools in the CNC and hasn't locked it out?"

Here's where all that previous relationship building comes into play. Even if this is the first time you've met Ken, this softer approach will get better results than writing him a safety ticket. The safety professional approaches Ken, sees he isn't wearing his safety glasses or hasn't locked out the machine when he should have and says in a friendly tone, "Hey Ken, can I grab you for a second? This is important." When Ken approaches, the safety professional says something like "I noticed you aren't wearing your safety glasses," or "It looks like that should be locked out right now," and gives Ken an opportunity to reply.

The safety professional is going to get some important information here. It could range anywhere between a sheepish, "Oh, yeah" to a more brutish "Go f- yourself." In almost every case, the safety professional's response should be about the same. "I really don't want you to get hurt, but I'm sure there's a reason you are doing it that way. Can you tell me why so that we can figure out a way to do this safely?"

That sounds strikingly similar to collaboration. There was no threat of discipline, even if Ken was a jerk about it. Also, did you notice the emphasis on *why* in that interaction? Rather than simply issue a command, the safety professional can get to the root cause of bad behavior just like we can get to the root cause of an incident by asking "why" questions.

"Why didn't you lock the machine out?" asked in a non-aggressive manner that conveys curiosity and not frustration often results in an answer something like "There's an interlock that shuts down the machine," or "I need to jog the head after each piece is removed, then jog it again after each new piece is mounted." Sometimes you will still get an answer in the neighborhood of "Didn't I just tell you to go f- yourself?"

---

*Control-reliable*
Simplified, control-reliable means that safety systems shut down a machine and put it into safe mode when a guard or other safety feature is removed or opened.

---

Once you've identified the root cause of the behavior, you've got a safety project on your hands. Can we design the interlock to be control-reliable so that lockout isn't needed? Can we find a better way to change the tooling on the head so that it doesn't need to be jogged so much? Do we need to talk to Ken's manager about anger management?

---

*Jogging a machine*
Jogging means very slowly moving the machine a tiny bit at a time to position it precisely or to observe its movements. Many machines have a specific jog button or other type of jog control feature.

---

To assume that addressing the root cause of the behavior is always this easy is silly. You and I both know that there are going to be some serious curveballs here. But by using a root cause methodology when an undesirable behavior is identified, we're going to get better long-term solutions.

*Who gets the citation?*
If OSHA issues a citation, it is issued exclusively to the employer. OSHA cannot issue a citation to an employee. Conversely, MSHA, the Mine Safety and Health Administration, can issue citations to individual employees under certain circumstances.

It's time to relate a story. This situation occurred when I worked at a shoe manufacturing facility. There was a machine that had to be disassembled and shipped out of the plant to another location. The machine had two metal columns on each side of it that were roughly twenty feet tall, composed of two sections bolted together. The columns had to be unbolted, taken down from top to bottom, and trucked out of the room via forklift.

*Working at height*
OSHA general industry standards require fall protection if an employee can fall more than four feet (29 CFR 1910.23). The walking and working surface standards will also preclude the practice of standing on an elevated pallet.

I was making my rounds on a Friday after the last shift of the week had finished, visiting the different areas of the plant. When I walked into the room where this machine was being disassembled, here is what I saw: Six maintenance employees were working on the project. One of them was standing on a pallet, elevated about nine feet above the ground on the forks of a lift truck. The top portion of the column had been unbolted, and the employee on the pallet was shimmying the column onto the pallet, walking backward as he did this. No fall protection.

I immediately pictured the employee stepping off the pallet as he moved backward with the huge metal column, the column crashing down onto his body after he slammed into the concrete. I spoke up, and said "Hey, guys, you gotta stop this. This is way too dangerous."

The man on the pallet stopped, looked at me, and said "F- you, a-hole," then went back to work. He was usually gruff, but not usually *that* gruff.

I swallowed back any alpha male aggression, and responded with "No. Seriously, this situation goes wrong and you can die. We need to find a better way to do this, guys."

At that point, I had all six of them telling me where to go, how to get there, and which OSHA standards I could shove into which body cavities on my way. Momma didn't raise no fool. I left the room to visit their boss.

I explained the situation to him, and his eyes grew large. "We need to get down there," he said.

The trip to his office and back took between five and ten minutes, and the entire time I was waiting to hear a huge crash echo through the facility, followed by horrid screams. Thankfully, I never heard those sounds. By the time we got back to the room, the top portion of the column had been successfully (but not safely) removed, and was being secured to a shipping pallet.

*Toolbox talk*
A toolbox talk, also known as a tailgate talk in construction, is a brief, informal review of the hazards of an upcoming task. Toolbox talks usually only take a few minutes; they should be collaborative as the group works out how best to safely do the job.

The maintenance department supervisor and the group got into a pretty good argument, which I simply observed. Like I said, Momma didn't raise no fool, and I knew it was a good time to stand back, shut up, listen, and learn.

What I learned was that the maintenance crew didn't know how else to take the machine apart. They were upset that they had to work late on a Friday, and this job was looking like it would take a few more hours. Finally, they really didn't like me because I always seemed to slow them down with all the stupid safety training, toolbox talks, pre-work huddles when there was a non-routine task, etc. When things had cooled down a bit, the eight of us talked about how to safely remove the next column. We came up with a plan to unbolt it while standing in a personnel lift basket, then tie off the top portion to the forklift; the forklift would then lower it to the ground.

*Root cause*
Mark Paradies, a well-known systems specialist, defines root cause as "The most basic cause (or causes) that can reasonably be identified that management has control to fix and, when fixed, will prevent (or significantly reduce the likelihood of) the problem's recurrence." Using this definition, the idea of finding a root cause can be applied to solving any problem.

The maintenance supervisor and I sat in his office when the safety-sensitive work was finished. "What do you want me to do about this?" he asked.

"I understand why those guys were upset, and I get that they didn't know what else to do. But their refusal to stop work when ordered and the lack of common decency are not okay," I said.

"Yeah," he said. "I'm going to have to write them up."

"Mike," I said, "do what you have to do as their supervisor. But can I get some time with everyone at your next staff meeting?" I wanted to get to the root cause not just of the near-hit, but of the poor attitudes. Mike had an obligation as the supervisor to write warnings to people, but from a safety perspective, I wanted to have a completely open, honest, and candid conversation.

I knew that Mike had a Monday morning meeting each week, and it was going to be a perfect opportunity to address the situation soon after it happened, but with enough time passed that everyone should be cooled down a bit. I needed the cooling off as much as anyone. Part of me wanted to fire the entire crew, but to be productive, I needed to take a step back and look at the bigger picture.

> *Not your job*
> As a safety professional, it is not your job to keep everyone safe. It is your job to give them the resources and environment they need to work safely. Each individual employee is ultimately the only one who can keep him- or herself safe. This can be a difficult concept for some people to understand, and I often try to reinforce to employees the fact that ultimately they are the only ones who can keep themselves safe.

When I arrived for the Monday morning meeting, it was a bit tense at first. None of the men on the crew offered an apology to me at any point, nor did I ask for one. I sat with them in the meeting room and said "We need to talk a little bit about Friday, guys. That was a seriously unsafe act, and when I tried to stop it there was a pretty bad reaction."

> *Rigging*
> Rigging means lifting and/or moving big, heavy, complicated stuff. It is a specialty, and rigging companies should be hired for tasks that involve complicated lifting and moving of machinery.

More talk from them about slowing things down was sent in my direction. To be honest, it sounded an awful lot like whining.

"What would this meeting be about today if you had fallen off the pallet?" I asked. Silence and a lot of looks at the floor followed. "Look," I continued, "I'm not here to chew anyone out. It happened, it's done, and I don't care if you don't like me. Personally, I don't like many of you at this point, either. But I do have an obligation to provide you with everything you need to be safe. So let's talk a little bit about how we can do odd jobs like this safely next time." From there we performed what was basically a debriefing of the incident. The future plan: Hire out jobs like this to a rigging company. The cost of overtime compared to outsourcing was basically a wash.

> *Debriefing*
> A debriefing is a meeting held after an event to discuss what went right or wrong, and what can be learned. In the Army, we called it an "after action review."

I could have approached the situation much differently. Part of me wanted to. I had to keep a long-term view, though. I could have reasonably asked for harsh discipline based on the safety aspects of what happened. Everyone would have understood; but I wouldn't have been given the same answers by the crew when we sat down to talk it out. We would never have gotten to a better long term solution, and future productive conversations would not be possible. Instead of building a wall, we worked together, grudgingly, to create a better relationship.

There are other times when a safety violation *should* result in discipline, but those situations should be few and far between, and based on severe violations that could result in death. Yes, you could argue that the above situation met those criteria, and maybe one or two people should have been fired. I wouldn't have been sad to see them go, but it would have been unfair to them, since they were simply ignorant of a safer alternative method to do the work.

---

*Firing people*
Unless you are supervising employees who are your direct reports, you should never fire anyone. Leave that to their supervisory chain of command. Don't threaten to fire anyone; if they end up keeping their job, you will look even worse than threatening them made you look.

---

Here's a better example of when safety discipline is appropriate: When I worked at the grain handling company, a facility manager was working on top of a grain silo that stood about thirty feet high. He needed to get to the next silo over, which was the same height and spaced about two feet away. Rather than climb down the ladder of the silo he was on, walk over to the other silo, and then climb up that one's ladder, he simply jumped the two feet of space between them.

---

*Catwalks*
A catwalk is a steel walkway that is elevated and gives employees access to areas that are otherwise difficult or unsafe to reach.

---

One of the safety team witnessed this, and immediately called the regional director, who was in the area. The regional director drove to the elevator and fired the manager on the spot. He called me afterward to ask if I thought he had done the right thing. I told him one of my mantras, "I'd rather see someone in the unemployment line than in a casket."

---

*Workers' compensation costs*
It is odd, but in most cases it is cheaper to kill an employee than to permanently injure them. In some states, the payments for death are laughably low. Having experienced a fatality situation in North Dakota, I can tell you that the work comp system there is a joke and a serious disservice to the wonderful, hardworking people of the state.

---

What are the differences between the situations? In the first, the employees had no idea how to mitigate the risk, and were performing their work in what they felt was the safest possible way. The employees on the floor below the man on the pallet were spotting him, telling him where he was on the pallet, and keeping a keen eye on the column. Terrible safety, yes, but an attempt at safety. The grain facility manager knew what the safer alternative was; he even knew that a catwalk should be installed as a long-term fix. Had the grain facility manager requested money to construct a catwalk between the silos, his request would have been accepted; all capital requests for safety were accepted at that company. A short catwalk and the contract labor to install it are cheap. Way cheaper than the death of an employee (except in North Dakota, where it is only marginally cheaper; the work comp system there is really screwed up).

> *ANSI*
> ANSI is the American National Standards Institute. They, together with ASSE write many of the voluntary safety standards that go above and beyond OSHA. When I write safety programs, I generally rely more on ANSI standards than OSHA standards. They are more well-written and up-to-date, and usually provide a higher level of protection.

Nearly every ANSI standard seems to have a section that requires discipline for failing to follow the rules. As a result, companies that try to go above and beyond OSHA standards by using ANSI for guidance shoot themselves in the foot. Trying to scare employees into compliance by threatening them is akin to the old saying "The beatings will continue until morale improves." It's time to embrace the idea that discipline is detrimental to the safety program in almost every case. Discipline should be reserved for those cases when an employee will be seriously hurt or killed and they are (or should be) aware of a safer method, or when an employee shows repeated blatant disregard for the rules.

The bargain between a safety professional and a violating employee works best when it is this: The EHS department will consider the violation a near-hit, get to the root cause of why the violation occurred, and try to solve the root cause of the issue; there will be a written record but only a warning, as long as the employee agrees to follow the safety rule as-is until we get the causation issue mitigated. Include the employee in the investigation of how to improve the process, and keep them updated on the progress. Most of the time there is a solution to be found; sometimes it is going to take forever, and other times, it will be a dead-end and the employee is just going to have to accept that they are stuck following a safety rule they don't like. By the safety professional offering up meaningful, goal-oriented solutions that focus on how the employee wants to work, the employee is going to be a lot more likely to comply consistently. People don't come to work to get hurt; if they are doing the job in an unsafe manner, they have a reason why.

After the situation at the shoe factory, the maintenance crew made efforts to be a little more open with me because I was accepting of their input and looked for solutions to problems rather than just bringing down a safety hammer. I was worried they would think that my lack of angry response was a sign of timidity, but I am pretty sure they could see how much I was holding back on my response to their disrespect. We still had challenges, and there were still times when we clashed; but I began attending every one of their staff meetings and eventually was viewed as an integrated part of the group.

It all came back to relationship building. I truly did not like some of the crew, and some of those same people truly did not like me. But we found ways to relate to each other, and many of the maintenance crew who I did get along with were also friends with those who tended not to enjoy my company (or very existence in one or two cases). The time investment I made built enough bridges that we could discuss things productively.

# 21 Here I am, stuck in the middle with you

It is a terrible, ugly, and at times chilling fact that our profession is directly affected by… ⋆sigh⋆ *politicians*. They're just awful, really, with only a few exceptions. I've met some politicians. A small number of them are genuinely nice (or incredibly good at faking it), and a few on the other end of the personality spectrum have zero ability to relate to everyday, sane human beings. Most are just on one side or the other of the middle of that spectrum. As a safety professional, I try to keep an eye on what politicians are doing and how it might affect my career.

I'm going to try to approach this subject in a manner that is as unbiased as possible, but it's going to be difficult, because of the two major parties in America, one generally tends to lean toward laws that support the safety profession, and the other tends to lean away from that support.

It's a sad truth that as safety professionals, politics plays a role in our jobs. OSHA, EPA, and other workplace regulators cause people's political hackles to raise. On one side, we've got those who say that workplace regulations don't go nearly far enough in their scope, and they argue that penalties must be stiffened and rules made more stringent. On the other side, we've got those who want to scrap regulatory agencies all together and let business regulate itself. I am going to try and be neutral but will probably present some arguments in this chapter that may not seem very centrist to you. Please keep an open mind.

When you are talking to employees, managers, or pretty much anyone you encounter at work, it pays to be as centrist as possible. If you are too far to either side, you will lose the ability to reason with the extremists on the opposite side. The right-wingers will see you as a job-killing communist who thinks that we should all walk around with hockey helmets on; or the left-wingers will think you are a heartless drone, pre-programmed by the soulless leaders in the C-suite to enslave us all in the holy name of Profit.

The truth is almost always somewhere in the middle, and to really allow a safety program to flourish it pays to have the ability to tactically explain to one person that the reason we have all these guards and signs isn't because we live in a nanny state or that kids nowadays are too stupid to live, but because human error is a normal part of life; then go and explain to another person that the company does not have endless resources, and we have to carefully plan how and when we address hazards rather than wave our magic safety wand and fix the world with no regard to budgets or the bottom line.

Thomas Frank, who started as a conservative journalist and has (as of this writing) changed his viewpoint to be more liberal, said that "The great fear that hung over the business community in the 1970s was death by regulation, and the great goal of the conservative movement, as it rose to triumph in the 1980s, was to remove that threat—to keep OSHA, the EPA, and the FTC from choking off entrepreneurship with their infernal meddling in the marketplace." This is still a common view among conservatives; that OSHA and other agencies are just getting in the way of business. Too many people forget history too easily. Prior to modern labor laws, anyone hurt at work was on their own. Children were not in school, they were in the workplace. And taking steps to make sure the environment was protected? Not a chance.

If you think that we could roll back regulations today and still have a world that looks and feels anything like it does right now, think again. Business is in business for money. Visit the factories in unregulated countries and see for yourself what work could be like here at home if regulations were gone. The "infernal meddling" that people complain about is nothing more than a method of providing us all with reasonable protections. If OSHA disappeared today, most of us in the safety profession would be searching for a job tomorrow. We wouldn't have to look too long though, because the high death and dismemberment rate would keep plenty of manual labor job openings popping up.

When conservatives argue that new safety laws and environmental regulations are "job killing," I have to scoff. I have a job *because of* safety and environmental laws. Requiring power plants to run in a cleaner, more efficient manner creates a lot of jobs for people to design, build, transport, and install scrubbers and other equipment; then keep that equipment running. Want to keep an economy humming? Keep people busy working and innovating. The job creators aren't the people who have billions of dollars in the bank, they are the people who have ideas to make work safer, cleaner, and more efficient.

---

*The OSHA log*
Employers are required to record workplace injuries (other than first aid incidents) on a document called the OSHA 300 log. Annually, a summary must be posted. This summary is the 300A form.

---

On the flip side, we have some examples of OSHA regulating beyond what are reasonable bounds. The new OSHA rule to electronically post company injury and illness records for public view (see www.osha.gov/recordkeeping/index.html for details) will not produce any positive safety results. By focusing on lagging indicators that members of the general public may not fully understand, OSHA is providing employers one more reason not to record employee injuries and illnesses. Employers become divided into two groups: those who choose to accurately report injury and illness data and accept the consequences of that reporting; and employers who lie. Potential employees, unions, media researchers, and others will be given a false view of the two groups, resulting in inaccurate evaluations of employers. Those employers who withhold accurate data will be rewarded by a lack of negative attention and a deep well of misinformed members of the public, all based on the perception that they do not have a high injury and illness rate. Employers who accurately report will oftentimes need to expend time and resources defending themselves against the negative impact of their honest numbers. Safety professionals' time will then be shifted from a focus on driving leading indicators to a focus on defense against false perceptions, and eventually most employers who begin with an intent to accurately report will be tempted to move toward inaccurate reporting. Let me be clear: This rule encourages and rewards the use of misleading, inaccurate information (aka lying).

OSHA has amended the rule with a three-step approach. None of the three steps will make any difference in the real consequences of the problems posed by electronic reporting. Proposal one is to require employers to inform employees of their right to report injuries and illnesses. This is already a regulatory requirement found in 29 CFR 1904.35. If an employer is not currently meeting the requirements of this standard, they will not do so in the future if electronic reporting is required. Proposal two is to more clearly communicate the requirement that any injury and illness reporting requirements established by an employer be reasonable and not unduly burdensome; as with proposal one, creating more language in an electronic reporting requirement will make no difference in an employer's decision of whether or not to accurately report. Proposal three is to provide OSHA an additional remedy to prohibit employers from taking adverse actions against employees for reporting injuries and illnesses. According to the OSHA website there is one compliance officer for every 59,000 workers. Does anybody really think that it is a good idea to take compliance officer hours away from worksite inspections to chase after employers who are avoiding accurate reporting into an electronic system? I know I would rather have the limited time of OSHA compliance officers dedicated to making workplaces safer through identification of hazards than being bogged down with trying to identify paperwork violations.

Finally, this rule is awful public relations for an agency that needs a PR "win." This regulation frightens employees into thinking that their privacy is at risk; it will further animosity toward government regulations from businesses that already feel overburdened; most chillingly, it will sour the relationship between many safety professionals in the private industry and the agency. For private industry safety professionals, this rule contradicts all the work we have done, oftentimes in coordination with OSHA, to break free of the reliance on lagging indicators to measure safety. OSHA could be changing safety standards to make them easier for the general employee population to understand, or updating permissible exposure limits of chemicals. Instead, they have taken a misguided approach to try and play the "shame game." The only ones who should be ashamed in this situation are Dr. Michaels and those around him, for allowing this rulemaking proposal to ever see the light of day.

So here we are with life in the middle. The simple facts are that we need OSHA, EPA, and other regulatory agencies, or we would be taken advantage of in serious and dangerous ways by the people who have the most power and money. On the other hand, we need to control the regulators in order to keep regulations relevant and the economy viable. Truth exists on both sides, but on neither extreme. In *Nicomachean Ethics*, Aristotle talked about a virtue as a middle ground, not exactly in the center of two points but near the center. If your safety philosophy becomes a virtue, you will win the battle of hearts and minds that truly makes a safety program successful.

If I could go back in time, I'd love to meet Francis Perkins. She was an amazing woman. The first woman to be appointed to a cabinet post, she was the Secretary of Labor for Franklin D. Roosevelt (I like to think they'd say good morning in the hallways of the White House by calling each other "Franky" and chuckling). Perkins helped create legislation to outlaw child labor, established the first minimum wage laws, and was one of only two cabinet members to serve under Roosevelt for the entirety of his presidency (Coleman, 2010).

---

*The triangle shirtwaist fire*
On March 25, 1911, a fire at The Triangle Shirtwaist Factory in New York City killed 146 workers, most of them young women. The fire started in a scrap bin on the 8th floor. The owners of the company had locked doors to exits and stairways, trapping the workers inside. Dozens of the women killed jumped or fell to their death as the flames advanced (Stein, 1961).

---

The biggest reasons I mention Francis Perkins here are because she also created the Bureau of Labor Standards in 1934 to promote safety and health in the workplace, and was instrumental in passing safety reforms after the Triangle Shirtwaist fire (she was appointed as executive secretary of the commission investigating the fire at the behest of Teddy Roosevelt). Perkins was already a recognized safety expert when she witnessed the Triangle Shirtwaist fire firsthand, and the tragedy motivated her to do more (Coleman, 2010). It also motivated a group of sixty-two people to found the United Association of Casualty Inspectors, which went on to become the American Society of Safety Engineers (American Society of Safety Engineers, n.d.). This period and the events that occurred as part of and coinciding with Perkins' leadership were the beginning of real change in American views of workplace safety.

To really get an idea of how safety and politics are tied together like two cats hung by their tails from a clothes line, we've got to go back even farther to some really shameful days of American history. Twelve presidents of the United States owned slaves, eight of them while serving as president (*Which U.S. Presidents Owned Slaves?* n.d.). One wonders how Thomas Jefferson could have written the words "We hold these truths to be self-evident, that all men are created equal, that they are endowed by their Creator with certain unalienable Rights, that among these are Life, Liberty, and the pursuit of Happiness" while owning slaves. To be fair, Jefferson wanted slow, practical abolition of slavery even though he would own slaves his entire life. Jefferson, Washington, and several other founders who owned slaves hoped and reasonably expected that slavery would fade as the cost of housing, feeding, and "caring" for slaves rose. This was before the cotton gin made slavery economically appealing, mind you (Meachum, 2013). In a society where slave ownership was acceptable, there was no realistic chance that strong workplace safety protections would be enacted.

The federal government did pass a few limited safety laws prior to abolition as far back as the 1790s. The small number of laws were aimed at shipping and rail, but there was no general duty of employers to provide a workplace free of recognized hazards (MacLaury, n.d.). Safety at work just hadn't become a priority yet.

Certainly, there were stronger voices for abolition than those who hoped it would go away on its own, but it ultimately took a war to end the practice of slavery. Those who opposed abolition used the arguments that slavery had always existed in the world, that without slavery the Southern economy, based first on tobacco and later on cotton, would be ruined, and that slave owners took better care of slaves than the slaves could take of themselves (Meachum, 2013; Axelrod, 2011). These arguments are reflective of today's oft-stated arguments against government regulation of the workplace: Resistance to change simply because it is change, the legislation will be "job killing," the economy will suffer, and those who are rich and powerful have levels of wisdom beyond those of the common man or woman.

Okay, just settle down and take a deep breath. I'm not trying to pick on conservative politicians, but the fact is that they've been using these same arguments, with minor word-changes, for over 150 years. What these politicians fail to see is that change is a natural, inevitable, and positive course in human events; regulation that provides reasonable protections to workers will be a benefit to the economy; and just because someone was either born into power or had the luck to become rich and powerful without the help of "old money," that doesn't mean they are any wiser than the rest of us.

Let's return to history. Remember when I mentioned that Congress passed laws governing safety on the seas and railroads dating back to the 1790s? Most of these laws started out pretty ineffective, but were strengthened over time. States took over and created a few laws as well, but politicians really ramped up the regulatory action after the Civil War. No, it is not a coincidence that after we freed the slaves, we decided that all workers deserved dignity and protection. The Department of Labor was created in 1913, and then we meet back up with Francis Perkins a couple decades later (MacLaury, n.d.).

What were some of the other arguments against workplace safety improvement laws during Perkins's time? According to the Reverend Edgar Gardner Murphy, who was a leader in ending child labor, the arguments businesses had used against him included "Do you not know that the cotton factories are the agents of prosperity?" "Do you want to compromise or to arrest the prosperity of the south?" and "Do you not know that this child-labor law is an attack upon business?" (The Social Welfare History Project, n.d.) Does any of this sound familiar? It should, because these are the same arguments that were used to justify slavery and are used today to try to curtail regulation.

On February 15, 2011, a House Subcommittee on Workforce Protections held a hearing on OSHA's regulatory agenda. When it came to strengthening the hearing protection standards, the National Association of Manufacturers argued:

1    The proposed OSHA noise interpretation would affect a large number and very broad range of American businesses and their employees.
2    The costs for American businesses to comply with OSHA's proposed new policy would be very high.
3    OSHA's proposed new interpretation would have substantial negative impacts on U.S. jobs and competitiveness.
4    All this would be for relatively little benefit in terms of improved hearing protection for workers.

At the same hearing, the Chamber of Commerce argued that OSHA's proposals "reflect a troubling pattern of efforts by the agency to impose substantial burdens on American businesses without regard to the cost of those efforts" (U.S. Government Printing Office, 2011). The arguments by the Chamber of Commerce and the National Association of Manufacturers are the *exact same* arguments that have been made against worker safety and health since slavery!

In 1967, President Johnson proposed federal workplace safety regulations, but the legislation was defeated in 1968. President Nixon tried again in 1969, and the result was the Occupational Safety and Health Act, which established OSHA. Politics being what it is, each party wanted *their* version of a federal workplace safety bill passed. The Act ultimately was a combination of the two proposals, and all sides seemed happy with it. This was back when compromise was possible between the parties (MacLaury, n.d.).

Many of the people who oppose federal safety regulations say that industry will self-regulate if all those burdensome OSHA standards go away. Am I the only one who sees the problem here? The same people who say that these safety standards are job-killing, that they hurt American companies when competing against foreign companies, etc., etc. are the same ones who promise to self-regulate! How is this not utterly laughable on every level? Since passage of the OSH Act, workplace fatality rates in the USA have declined by over 60 percent (OSHA, 2012). If self-regulation works so well, why were the rates as high as they were pre-OSHA? The simple fact is this: Without the OSHA python, companies would not provide nearly the level of workplace safety to employees they do right now.

The real reason that self-regulation is so desirable is that it allows companies to shuffle safety to a lower priority at will. Companies get a nice win–win with this. They can talk about how much effort they put into self-regulation and how safety is always named as a top priority, but they can simply ignore, defund, or otherwise abandon safety as needed. And when will "as needed" happen? Almost constantly. Forget about a decent salary and continuing to increase the professionalism of safety; we'd be back to the safety job being reserved for the person who got hurt too badly to do his regular job, or for the owner's kid who is a complete schmuck and will do the least damage in the safety office. Self-regulation allows management to see safety the way I see a sink full of dirty dishes: Ugh, what a pain in the butt. I'm just going to leave this here until someone else happens along and gets stuck taking care of it.

---

*Professionalism*

Please don't be offended if you are a person who was hurt too badly to do your regular job and were "promoted" to the safety manager position. If you are the owner's kid, you aren't a complete schmuck, because you're reading this book and trying to make yourself into a more polished safety professional. If you wear the safety hat along with your eighty-seven other duties or have otherwise "landed" in the safety management chair, please continue reading books like this one and learning more about the safety profession every day. Just because you may not have a background or education in safety, that doesn't mean you can't be a great safety professional.

---

How many times has an employer asked you (if you are a safety professional) to show them where OSHA requires doing something? How many times have you made the case for going above and beyond an OSHA standard only to hear management tell you that if they are not required by law, they won't take the action asked by you? Now think about how the people who made those statements would handle safety if there was no OSHA.

It is time to stand up and stop accepting the same excuses that have spewed from the mouths of politicians from the days of slavery onward. Safety regulations (and for that matter, environmental regulations) do not "kill jobs." What is the first thing that would "kill" your job as a safety professional? Obliterating OSHA and the EPA. What do the politicians who want to slash and burn government regulation want to do? They want to "kill" **YOUR** job!

# 22  Lift with your head

Forget everything you've learned (and taught employees) about bending at the knees and lifting with your legs. According to the Bureau of Labor Statistics website, back injuries account for one out of every five workplace injuries or illnesses. The American Chiropractic Association claims that 31 million Americans experience low-back pain at any given time (American Chriopractic Association, n.d.). If you aren't a statistics person, think about back injuries like this: how many people do you know who at some time have had back pain? How many people do you know who visit a doctor or chiropractor for back pain on a regular basis? How many have had surgery for their back pain? Most of the people you just thought about have been trained over and over again to bend at the knees and lift with their legs, yet they still had a back injury.

Nobody really knows when safety training began to embrace the idea of teaching people how to bend at the knees and lift with their legs, but it has been taught for many, many years. I picture a Caveman Supervisor grunting out instructions to a new Junior Rockpiler about it after singing the Cave 74 anthem to start the workday. Many people are so used to this mantra of bending at the knees and not using their back that they can recite it from memory. So if the concept of lifting with the legs is so well known, how come people still lift incorrectly and hurt their backs?

To answer this question, shift the way you think about an individual human being and picture us as carefully engineered machines. We come pre-programmed to do certain things certain ways. If the way we do those things provides us with a successful outcome, our processing unit logs the success, and we are more likely to *unconsciously* do those things the same way in the future; in this way, our "programming" learns what works most efficiently and effectively to accomplish our goals. When it comes to lifting, our programming started when we were very young. We learned that if we bent our knees as far as they will go and dipped our body low to the ground, we lost a lot of the mechanical advantage that is built into the levers created by our knees. So we added a series of other levers by bending our back. The several muscles in our back, combined with less of a bend in the knees, made for a successful lift. The result is that we learned to lift "incorrectly."

By the time we enter the workforce, we've already been so strongly programmed to lift like this that many people *think* they are lifting with their legs, when in reality, their back is doing a large portion of the work. In addition, the incorrect lifting we've done over the years has already begun to damage our backs. The commonly repeated anecdote about the person who throws out his back bending over to pick up a pencil happens not from that single improper lift, but from all of the improper lifts in that person's history causing small amounts of damage over and over. This was just the pencil that broke the rockpiler's back. This same concept applies to nearly every back injury; the injured person's entire history leading up to the injury helped "prepare" their back for the single fateful day when the injury occurs, because the discs of the back do not receive blood flow. With no blood flow, the discs are unable to heal themselves like the rest of our body, and chronic tears grow worse as multiple small areas of damage merge and grow (Spine-health, n.d.). As an analogy, think of back injuries like heart attacks. If I have a heart attack while eating a greasy cheeseburger, it wasn't that particular cheeseburger that caused my heart attack, it was the tens of thousands of cheeseburgers I've eaten prior to that one, all adding up to clog my arteries (and yes, I admit that I eat way too many cheeseburgers). I'll gladly pay you Tuesday for an arterial stint today.

---

*Revisiting the hierarchy of controls*
As a quick reminder, according to the hierarchy of controls, eliminating the hazard through design and engineering is the best mitigation option. Teaching people how to lift is an administrative control, which is less preferable. There is no personal protective equipment for lifting. Back belts are evil, throw them in a big pile in the parking lot and light them on fire.

---

If bending at the knees and lifting with the legs isn't the right answer, then what is? I argue that the right answer is to lift with your head. Take a look at what needs to be lifted, hefted, moved, shifted, etc. and then think about a way to use a tool to do the work. Tool use is natural to us as humans, we've been doing it since before Homo sapiens even existed (Swisher, Curtis, & Lewin, 2001). Hand trucks, dollies, carts, buckets, and other tools remove or greatly reduce the forces placed on your body, thereby reducing the chance of a back injury. In safety, the first way to prevent an injury is to eliminate the hazard. When it comes to lifting, this means finding a way to use a tool to bare the weight of the lift. Think about tool use for lifting like this: If you don't want to hurt your body lifting something heavy, don't lift something heavy. It's just like my fear of being bitten by an alligator; to keep that from happening, I don't wrestle them.

---

*Reviewing loss runs*
Your company's work comp carrier can provide you with a loss run, which is a snapshot of the work comp claims your company has had over a given time. It is a good idea to review these and get a feel for what types of injuries are being reported.

When examining job tasks that require lifting, you have a great opportunity for experiential learning with employees. Use your trusty Gemba walk techniques and start to learn the job. Experienced employees will have come up with their own "cheats" for the heaviest work. Use what they've come up with to make the job easier, and build from it. If they slide the item across the floor rather than lift and carry it, try to find a device they can roll it across the floor on. If they keep a sack on a table and tilt it to pour material into a hopper, find or develop a tool that can both lift the bag and tilt it. Work with your engineering resources and commit time to browsing for product ideas on the internet. After you have gained experience with solving a few lifting riddles, you should be able to provide employees, designers, and other parties with suggestions based on past successes. Don't be afraid to mechanize a lifting task. A review of your company's workers' compensation claims history will give you the economic cost of back injuries (remember to calculate the indirect costs), and the math comparison between incurring more back injuries versus installation of a lifting system (that will also increase efficiency) should greatly aid in the decision-making process.

---

*Back injury costs*
According to the State of California Department of Industrial Relations, the average cost for a back injury is over $50,000 in direct costs (State of California, n.d.).

---

By now, you should be wondering "What about using two people to lift a heavy object?" Good thought. There are problems with the two-man lift, however. When I was in the Army Corps of Engineers, I learned how to build a really neat style of bridge called a "Baily bridge." They are modular and carried in pieces on trucks, so that a group of engineers can build them by hand. As the bridge is built, it is pushed out further and further over the gap, until it reaches the other side. They've been around since World War 2 and are still produced and used around the world. Comrades who served in Europe told me that a fair number of Baily bridges constructed during World War 2 are still in regular use. When I and my fellow trainees learned how to construct a Baily bridge, the drill sergeants (with the utmost kindness and respect, as is always the case in basic training) lined us up by height in teams of four. Each team of four soldiers was about the same height so that when we grabbed the bridge piece and lifted, no single soldier was lifting his piece higher than everyone else. This, the drill sergeants explained was because "If that happened, the tall lanky [gentleman] over there would bare all the weight, and you short, lazy [valued assets] would have it too easy." As we picked up the pieces and carted them by the drill sergeants, they would take a good look at all of us to see if someone wasn't carrying their portion of the load. If they saw a soldier who looked like he wasn't pulling his share of the weight, they'd kindly tell him, "Quit being a [Richard Dawson] and use your [lovely] muscles before I have the rest of these guys drop this [wonderful] bridge piece on your [handsome] foot, [young sir]!" My lessons from this experience are threefold, and they all illustrate why using a multi-person lift is not the answer to moving a heavy object:

1    Unless each individual is of the same height, the weight distribution is uneven (think of the physics of the lift as if each person is a crane with a different height mast, and fixed, equal cable lengths… The crane with a tall mast will lift its side of the object higher, shifting the center of gravity and making the lift immensely more difficult).
2    In any group, the commonly accepted social psychological effects of social loafing, the free rider theory, and the sucker effect (these are all completely real and scientifically accepted phenomenon) will come into play; therefore, the group members who are exerting more effort are now quite possibly more likely to have an injury than they otherwise would have been (Krumm, 2000).
3    An additional risk that becomes apparent given the points in number 2 is dropping a heavy object on someone's foot when the anticipated muscle expenditure is not provided.

Are you growling a bit, thinking about that one lift back in the Fraggle Department that nobody has been able to engineer out for the longest time? Or are you growling a bit because you've already looked up the work comp numbers and presented your sure fire lifting solution to management, only to have it shot down? I hear ya. Lifting solutions are some of the toughest nuts to crack. Don't be afraid to commit a lot of time to the problems you see with current practices and devices. Touch base with your safety peers, hire outside consultation expertise, video record the activity so you can review it at your desk with the ability to pause or watch the images in slow motion, and if you can't get funding for the engineering solution you want, bring in a loss control specialist from your work comp insurance company (they hate back claims). Lifting is one safety issue where you will need persistence, additional brain power from other resources, and a serious ability to think outside the box.

Bending at the knees and lifting with the legs is an old paradigm. It is time for something new that matches the way we look at injury prevention in the 21st century. When there is a heavy or challenging lift, teach employees to go get a tool, and lift with their head. By the way, current exoskeleton technology is on the verge of completely eliminating this risk (if you can afford to buy a suit).

# 23 Written in blood
## The Cocoanut Grove fire

On November 28, 1942, over a thousand people packed into a nightclub that was rated for a maximum capacity of 460 named The Cocoanut Grove in downtown Boston. The club was dark on the outside, because Boston was under blackout orders due to the war. Inside, however, it was full of light and lavishly decorated. The owner of the club, a man named Barney Welansky, had enough political connections not to worry about the fire code. Exits were blocked by decorations and furniture, and in some cases even bricked or boarded over to prevent "dine and dash" problems. The club was decorated in a tropical theme, with paper palm trees and satin canopies. Patrons entered and exited the crowded space through a revolving door (Esposito, 2006).

---

*Methyl chloride*
Methyl chloride is a highly flammable gas that was (but is no longer) used as a refrigerant.

---

Shortly after 10, a fire started on the lower level of the two story structure. It spread to every area of the building in minutes. The true cause of the fire is still unknown, but it is known that the fire spread very quickly because many of the decorations and building finishings were made of flammable materials (Esposito, 2006). Additionally, the kitchen used methyl chloride as a refrigerant due to a Freon shortage (Kenney, 1999). There are several hypotheses as to the exact cause of ignition, but no definitive cause has been, or likely ever will be, found.

The Boston Fire Department responded quickly, but the condition of the exits, the huge crowd, and the rapidity of the fire's spread prevented them from being fully effective. The revolving door became blocked almost immediately as panicking patrons tried running out through both sides and were then piled on from behind as others passed by exits to leave the building the same way they had come in, something that has been shown through research as a common behavior in fire situations. The exits that were not blocked opened inward, and were not effective as people ran to them and could not pull them open against the weight of others pushing forward to get out (Esposito, 2006).

The constituents of the fire load in the building were extremely toxic, and many victims died in their seats of smoke inhalation. As more fuel burned and released gases, the fire grew more dangerous and people who did not escape quickly never escaped at all, with the exception of a group of five survivors who hid in a walk-in refrigerator (Esposito, 2006).

By the time the fire was out, 492 people—32 more than the rated capacity of the building—had died. Investigations began almost immediately. The findings of these investigations led to changes in the fire code regarding revolving doors, emergency exits, exit lighting, and furnishings (Esposito, 2006). Check out the next revolving door you go through, it either folds up if people rush it from both sides, or it has outward-swinging doors adjacent to it (maybe it even has both features). If you've ever tried to exit a building and the door pulled inward, it probably threw you for a loop because every fire code calls for exit doors to swing outward. You can thank The Cocoanut Grove fire and the 1908 Iroquois Theater fire (602 reported dead) for the outward swing of public buildings' doors (Brandt, 2003).

Barney Welansky was convicted of 19 counts of manslaughter. I don't know why prosecutors decided on the number 19 instead of 492. He was pardoned after less than four years, and died of cancer shortly after being freed. When he was released, he told reporters "I wish I'd died with the others in the fire." Barney had ignored the fire code, putting the profits of his business above the safety of his employees and patrons. His political connections paid off right up to the end. Maurice Tobin, the Massachusetts governor who pardoned him, had been an ally when he was Mayor of Boston at the time of the fire (Esposito, 2006).

In developing countries, building code and fire code violations still kill high numbers of people. As these nations develop and the citizens there demand better codes and laws, opportunity exists for safety experts to provide important consultation services to foreign governments. Here in America, the National Fire Protection Association has been trying for years to persuade municipalities to add fire sprinklers to new residential construction requirements. New building materials and methods lead to quicker fire spread, creating greater hazards for those living in new construction, as well as the fire fighters who enter burning homes to save residents. Building contractors fight the proposed sprinkler requirements, citing the increased cost that will be added to new construction. When I have listened to contractors' arguments against home fire sprinklers, their numbers are inflated, and I have no idea why such resistance hangs on so tenaciously. The fire code, like all other safety standards, is written in blood. I fear that because hundreds of people cannot die in a single residential fire, most municipalities will never face enough popular pressure to make such a change. As long as building contractors can sleep at night knowing their resistance to the change in codes directly leads to the deaths of the men, women, and children who live or fight fires in the homes they've built, they will put up enough of a fight to stop the prevention of these deaths.

# 24 DAF

This chapter is absolutely going to offend many of you...

It was a pineapple can lid. Melissa was making a homemade Hawaiian pizza, and my task was simple: Open the can of crushed pineapple, drain the juice, put the pineapple in a bowl, and throw the can away. No problem.

Opening the can went just fine. I drained the juice into a glass for a sweet treat later, then dumped the crushed pineapple into a bowl. So far, so good. Before me I had two items for the recycling bin: The pineapple can and the lid to the can. I usually place the lid inside the can, then drop the can into the plastic bin for recycling. This night, I entered the realm of DAF. I decided that it would be fun to bend the lid in half, just to show that I could do it. I held the lid like a platform, with my right thumb, pointer, and middle fingers acting as three legs, and I pushed straight down on the center of the lid with the pointer finger of my left hand.

> **NEISS**
> NEISS is the National Electronic Injury Surveillance System. It is a system set up to track injuries that are caused or contributed to by consumer products. If you go to the hospital because a consumer product has hurt you, your injury is entered into the database.

Now, since most of the dozen or so people who pick this book up and read it are going to be safety professionals, you're all picturing this and thinking to yourself, "This guy's an idiot!" And you're right. It was an idiot move. Physics never entered my mind. If it had, I would have realized that I was about to slice my hand wide open.

When I pressed down on the lid, it instantly turned ninety degrees and opened up the flesh between the thumb and pointer finger of my right hand. I looked at the wound for about 0.003 seconds before I realized that it was a deep cut. I could see meat.

Melissa gasped as I pinched the wound hard to both close it up and apply pressure with the fingers of my undamaged left hand, and thrust my hands straight over my head. Her eyes were wide and she asked, "How bad is it?"

I tried to smile reassuringly. "Honey," I said, "we're going to have to make a quick trip to the emergency room."

The color drained from her face and her eyes grew even wider. Before two minutes were up, I was calming her down and trying to keep her from getting frantic (still holding my hands above my head and pinching as hard as I could). About five minutes after that, we were out the door and on the way to the E.R.

The "quick trip" took about six hours or so, and when we got back, the cats had gotten onto the counters and eaten most of the pizza ingredients we had left out. Mable meowed at me when we walked in the door, and I swear she burped out the odor of white cheddar a second later. Edgrr's normally white muzzle was stained orange from pizza sauce.

While we tried to figure out what to eat now that the pizza was working its way through the cats' digestive tracts, Melissa asked me for the sixty-fourth time what I was thinking. I shrugged and said, "I don't know. I think that the Dumb Ass Factor got me." For those of you who don't speak French, you can refer to this as the "Darwin Award Factor."

I know what you're thinking, and you are absolutely right. As safety professionals, we should never say that someone got hurt because they were just plain stupid. But looking at this situation, there were no faults with the management system. After a few years of marriage, I had been properly trained and supervised. There was no need for PPE; when properly using a pineapple can, there is very little chance of getting cut with the lid and cut-resistant gloves are not necessary. The can had been engineered as well as it could be. There was no difficulty in removing the lid or defective part that led to my cut. It was just a really stupid move on my part, and in the real world, people sometimes just do really stupid things for no good reason. I like to call it DAF, and it is a major component of human error. As safety professionals, it is our duty to assume that any operator will have a "poor judgment day" and do something immensely stupid. It happens to us all, and most of the time we get away with it unscathed.

I analyzed the injury situation the next day as I admired my three stitches and picked up replacement ingredients for another attempt at a Hawaiian pizza. I knew that the lid of the pineapple can was razor sharp, and I knew that if it was mishandled it would cut me. Yet I still made the conscious decision to mishandle it and try to be impressive by bending it. If the workplace goal of a task had been to bend the lid and it was my job to perform a risk analysis for that task, I would have recommended using a tool. So when I wanted to bend the lid, why didn't I use a tool? Because I had become Cave Scott, and I wanted to show Melissa I was strong, I wanted to exert my power over the lid, and I had more testosterone in my brain fluid than logic and reason could override.

---

*Safety bingo*

There are a few ways to waste your safety budget worse than safety bingo. Don't make safety a game. For the most part, safety bingo has disappeared, but I still see a few misguided souls waste their budget on it.

So how does DAF play into the world of safety professionals? For starters, if Melissa had offered me a stamp on a bingo card if I had no injuries during the dinner hour or been observing me with a safety checklist, I would have still made the decision to bend the lid. If there was a poster hanging in the kitchen that said my safety was Melissa's number one priority, I still would have tried to bend the lid. As an animal, I want my reward (showing off to my wife and boosting my fragile male ego by bending a piece of flimsy steel) now, not later after thirty days of not hacking myself open with normally safe food packaging. Next, by offering a reward for a lack of injury, the implication would have been that I made a conscious decision to hurt myself. I did make the conscious decision to bend the lid, but the risk of the action never became recognizable while in the midst of DAF. The last problem with the bingo card or BBS proposition is that had I successfully bent the lid without an injury, I would have gotten rewarded with a stamp or nobody being there to observe me doing something stupid and being lucky enough to get away with it.

Next, I created an unanticipated situation by deciding to use the lid in a manner not foreseen by the pineapple can design engineers. They had designed a food package with an acceptable level of safety; in other words, when they evaluated the package it seemed to them as safe as it could probably get. There may even have been a warning label on the can that said something to the effect of "Don't try to impress your wife with this thing, it can cut you." It was my poor decision making, coupled with a lack of imagination by the design engineers to foresee such poor decision making that caused the incident. Had it been a workplace situation, some poor safety nerd would have had to try to complete an investigation and present it to upper management a) without laughing at me, and b) without calling me a total idiot (which I freely admit to being in that moment). What usually fails to occur is the next step of asking why the package wasn't a foil pouch instead of a can. There are no sharp edges with a foil pouch.

The lessons of the pineapple can lid for me were eye openers. In a production facility with four or five hundred employees, what are the myriad of factors that play into the decisions they make without even really thinking about those decisions? It is impossible to tell. When I look at safety solutions for dangerous tasks, I think a lot about the pineapple can lid (I kind of wish I would have saved it, but the scar is enough to remind me of its importance to my safety philosophy). People are going to stick their fingers in places they shouldn't for no good reason other than to see what happens. They will stick out their hand to try and catch a sharp object that falls off a work counter, and they will reach out and touch something that is obviously hot just to see how hot it feels. These things are all part of human fallibility and intellectual exploration.

---

*Rewarded for unsafe behavior*
Completion of a task is a reward. If an employee completes a task in an unsafe manner but happens not to get hurt, the employee has now been rewarded for working dangerously.

*Expanded steel*
Expanded steel is steel that has been stretched to leave diamond-shaped holes. It is an efficient and strong material that is often used for catwalks and machine guards.

DAF will never go away. It is immortal and as old as the first cave dweller who chucked a rock at a wasp nest just to see what would happen. Our job as safety professionals is to design systems that battle DAF, beginning with engineering that is as undefeatable as possible. The expanded steel box over the belt and pulley is our armor, it is our shield against the attacks of DAF; it is, indeed, our antithesis to the chaos of pineapple can lids on a high testosterone night.

DAF means that sometimes unexplainably poor decisions beat imagination, planning, and engineering. So what is a safety professional to do? We follow the hierarchy of controls and either substitute a less hazardous alternative or improve the engineering of the process; all while assuming that the operator will make the stupidest possible mistake. If the pineapple can had been a foil pouch, there would be no way I could get cut. If it were still a can, and I had to bend it in half as part of the disposal process, then this injury would be a perfect illustration of why I needed a tool for the job. Even though DAF is the root cause, we can design a work process that defeats it.

DAF is a mental error, oftentimes accentuated by distraction. A huge issue that is becoming apparent in safety is distraction due to devices. The more connected people are, the more their mental faculties will be focused on social connections. Millennials love their devices and staying connected to all their friends via social media. Our challenge as safety professionals is working around those behaviors in order to continue providing a safe workplace while not alienating the young workforce.

# 25  *To serve man*

I love classic episodes of *The Twilight Zone*. I think I've seen every one of them ten times, and they never get old. One of my favorites is titled *To Serve Man*, an episode from 1962. Interesting trivia fact: The alien in the episode is played by Richard Kiel, who went on to play the iconic character "Jaws" in the James Bond franchise.

If you aren't familiar with this episode of *The Twilight Zone*, it goes something like this [spoilers!]… A race of aliens lands on earth and starts solving the world's problems. Humans being humans, we trust them implicitly after they pull a few miracles out of their sleeves. They've ended the threat of war, so a group of code breakers from the Pentagon who now have nothing to do starts translating a book that belongs to the aliens but has been left behind at a United Nations meeting where the aliens gave a speech.

The code breakers first determine that the title of the book is *To Serve Man*. Here's where it gets *really* interesting. The aliens are giving people free rides back to their home planet, which they promise is a paradise. The leader of the code breaker group is about to board a ship to visit the aliens' home planet when one of his employees runs up to the ship.

"Mr. Chambers," she shouts, "don't get on that ship!"

*Hmm*, the audience wonders. *I wonder why she is so panicked.*

She continues, "The *rest* of the book *To Serve Man*… It's a cookbook!"

You bet your Betty Crockers that Mr. Chambers tries to run, but there is an alien there to stop him. He's forced into the ship, where the aliens fatten him up on the trip to their home planet. Mua ha haaaa!

How about this story for you… I had a friend of a friend (okay, it was a guy who stole a girlfriend from me years ago) who worked for a big food company here in Minnesota. He worked in the cereal division to be specific. He used to say that the company wasn't really a *cereal* company at all. It was really a *marketing* company that focused on marketing breakfast grains. The amount of money that went into polishing the image of a certain cartoon tiger (this person's department) was huge. Yes, the company made all sorts of food, and yes they spent all sorts of time and money on food research, finding new recipes, producing and packaging food, and everything else that a food company does. But they made money because they thought of themselves as a marketing company. Their customers aren't buying the flakes that are coated in sugar, they are buying the tiger on the box.

I share the two stories above because safety professionals, like the aliens in *To Serve Man* and the giant cereal company with the cartoon tiger, aren't exactly what we seem and our modern philosophy needs to change to reflect this fact. On the list of all the things we *really* are, the top spot isn't held by titles like "engineer" or "technical specialist." No! Our number one role is Customer Service.

Whoa. I know. It surprised me, too. One of my employers, a manufacturer, was pushing the idea that everyone in the company has customers, and it is each individual's job to serve those customers in the best possible manner. I decided to take this philosophy to heart, and the acceptance of my safety programs grew immensely.

---

*Evaluating chemical hazards*
I'm no chemist. My favorite methods for evaluating the hazards of a chemical are to use the safety data sheet coupled with reading the chemical's wiki page.

---

When a department manager called me in to consult on a new chemical, I decided to pretend like I was an outside consultant who really needed the business I was getting that day. I smiled more, I made myself more approachable, and I made sure that I always presented people with multiple options. I gave them what I felt was the best option, with an explanation of why it was the best option, then I gave them other possible options, along with the reasons they weren't the best.

---

*Reports, charts, and you*
If you aren't computer-savvy, sign yourself up for a few classes. In this day and age, to be an effective safety professional you've got to be able to use a computer to write programs, create spreadsheets, build training presentations, and do about a million other tasks.

---

I offered to look things up for people, and presented my findings in more of a report-like manner, always focusing on the task as if I needed the business. When I was called to perform ergonomic evaluations, I thought about Laurie, who runs my favorite ice cream shop, and the way she greets people and spends time talking with each customer about ten things other than ice cream before the interaction is finished. I began spending more time on the ergonomic evaluations, asking employees who the people in the picture frames on their desk are. The conversations led to friendly discussions about life in general, and I started learning more about employees' habits and hobbies outside of work that might be part of their ergonomic issues.

Managers looked at my responses to their questions and commented on how much they liked the information and the way it was presented. They asked more questions, and came to me more often when they had safety concerns. Employees did the same, and my Gemba walks became more productive.

There's a tricky place in this new path of customer service, and that's the understanding that the customer is not always right. Not only that, but the customer can get themselves and others hurt badly or even killed if they don't take your advice. Success with a safety program really loops back to relationship building at this juncture. The understanding and trust that you build with your customers over time pays dividends. They'll go along with an idea they aren't crazy about because you've shown them through your actions as a customer service agent that your primary interest is to serve their needs.

To serve people's needs, the safety professional has to identify those needs. This part is relatively simple; just ask. Think of it as a sort of Gemba walk and if possible, couple the conversation with a tour of the department or facility. Walking while talking keeps the ideas flowing. Managers will give you a taste of what their challenges are in the department. Oftentimes, the conversation turns to their biggest safety concerns. As a good customer service agent, you will put the manager's concerns somewhere on your priority list. Line employees will share their concerns with you, and when you compare and contrast these conversations with the department manager's concerns, you will begin to get an idea about where there may be a need to build communication bridges.

Marry the customers' priorities with your own to create the various to-do lists for your area of influence. If the customers see you diligently addressing their priorities while you work on your own, it brings a sense of validation. Provide lots of updates and keep getting input as you work through your priority lists, and remember that while the customers' perceived needs are an important part of what you do, the customer is not the safety expert; the final decision on where items land on the priority list is yours.

Just like the wonderful customer service reps from vendors you work with to get your job done, it is important to create a "check in" schedule with your customers. Believe me, they are going to wonder what in the heck you are calling or stopping in for if there aren't any safety issues. But they will be happy to see you if they come up with a problem you can help them solve. Great practice for new safety professionals is to work a few years in the loss control department of an insurance company or a consulting firm. It forces us to work with real-life customers, and the skills learned there can be carried over into work inside a company.

As the new century takes firmer shape, more and more opportunities will arise that require safety professionals to use a customer service philosophy. The millennial generation seems odd to many of us who are established in the profession, but as the Baby Boomers retire out of the work force, my generation (Generation X) is going to be dealing more and more with people younger than us who have an entirely different outlook on life. Millennials expect to be served, and expect to be cared for by employers, insurers, and others in power positions (Twenge, 2006).

To retain Millennials in the workplace and to ensure they are a productive, active part of a safety program, a customer service approach is incredibly important. When I say that this generation seems odd to my generation, I know what I'm talking about—I am married to a Millennial. I'm only eight years older than Melissa, but there are significant differences in our generations. The internet, email, cell phones, satellite TV, and many other tech influences were either non-existent or in their infancy when I was a kid, while Melissa grew up with most of them. The computers she used at school made the ones I used at school look like Jacquard looms.

To keep up with Millennials, the safety profession has to understand that young people today (according to some researchers) have a stronger focus on their own wellbeing than previous generations. Whereas members of Generation X have on average a higher number of job changes throughout their career than Baby Boomers did, Millennials have even more job changes than Gen Xers (Twenge, 2006). I believe Millennials' job changes have reflected their philosophical upbringing in two ways: First, employers have a reputation for holding little to no loyalty to their employees. Like Generation X, Millennials have witnessed friends and family members be wronged, laid off, and generally treated in what they see as an unfair manner by employers. Second, frequent job changes are seen as a quick path to higher salaries and management positions.

Both of these motivations are justified and understandable. A Harvard Business School study found that CEO's make more than 350 times the salary of average employees. It is difficult for Millennials to justify sticking with a job that sends an unsatisfying paycheck every two weeks when the CEO is paid a king's ransom. According to the Washington Post, the average Fortune 500 CEO in the United States makes over $12 million per year (Ferdman, 2014). These are the same people who have laid off Millennials' parents, aunts, uncles, friends' parents, etc. If a CEO reduced his salary from $12 million to a paltry $6 million per year (barely enough for groceries, I say!), that would be enough payroll for sixty jobs with a total compensation package of $100,000 per year. Millennials understand that businesses have a moral obligation to provide jobs in our communities, and they also understand that CEOs are failing to meet that moral obligation.

Here's another way to look at it for those of us in safety. A well-paid safety professional with a $100,000 salary (don't we *wish* we could earn that!) is still earning way more than most unskilled employees, but the CEO still makes about 120 times the safety professional's salary. Millennials aren't stupid, and they can see that these giant CEO salaries are not just taking money away from their paychecks, but from the shareholder returns on their 401k packages.

All of this actually helps relieve some of our challenges in safety. Millennials are less likely to put their life and limb on the line for what they see as a soulless employer. Whereas Boomers and some Xers will take more chances to get a job done, Millennials are more likely to recognize that no job is worth their life. Many people of my generation and of the Baby Boom generation see Millennials as self-centered; I like to think that they are really just better at balancing their personal needs against what is reasonable or unreasonable of employers to ask of them. Like the aliens in *To Serve Man*, the cereal company, and safety professionals, they aren't what they seem. For safety professionals to make a real connection with them, we have to shift to a focus on their personal well-being. That doesn't mean we have to be their surrogate helicopter parents or give them a trophy every time they show up to a training event on time; it simply means we need to know our customers and serve them accordingly.

# 26 Snickergate

*Safety training sucks*

Make training fun. Nobody likes to sit through the annual mandatory training sessions. Skip using videos, and change training up each year to keep interest alive. Add a few fun things like trivia questions, funny YouTube clips, etc. and do what you can to get a high level of participation.

It was time for the annual all-employee safety meetings at the grain company. The schedule called for me to lead dozens of all-employee meetings in fourteen different states. It was going to be a grueling schedule, with out of town travel over 50 percent of the time from November to the middle of March. I would fly or drive to the first location early Monday morning, and sometimes host a Monday afternoon meeting. Tuesday through Thursday consisted of a three hour morning meeting, a drive to the next location, then another three hour meeting in the afternoon. The second meeting was followed by another drive, this time to the location of the next day's morning meeting. As the presenter, I had to pack a projector, my laptop, handouts, and other supplies.

The first meeting was in eastern Washington and my local contact, Wayne, a jovial guy with an infectious smile and a good sense of humor, picked me up at the airport. We would have the first meeting the next morning, and by the time we were done, I would have worked through Washington and Idaho, finally flying home from Great Falls Montana.

Wayne had seen the presentation, and he liked that I inserted some fun trivia questions every few slides to keep the audience engaged. While I set up the first meeting room, Wayne ran to a local grocery store and picked up a bag of bite-sized candy bars. "You should give these out as prizes when people get a trivia question right," he suggested.

I thought it was a great idea. The first meeting was small, with about twelve attendees. When the first trivia question came up and I told them they'd get a candy bar for the right answer, they all started shouting out their answers. I tossed a candy bar to each of them, not wanting to leave my space at the front of the room to gently hand them the bars. By the time the three hour long presentation was over, I was throwing a candy bar randomly into the crowd every minute or two.

The bigger the groups got, the more raucous the candy throwing became. In Montana, two employees started whipping their candy bars at each other as a distraction so the attacker would have a better chance at getting the trivia questions right. Nobody forgot that meeting, and I laughed so hard my ribs hurt.

Weeks swept by, and word of the candy bars spread. People were emailing me requests. "I love Mr. Goodbar," or "Mars products only, please." In a tiny town in North Dakota, when the presentation ended, the entire room of about sixty people threw candy bars back at me in one huge barrage. I laughed so hard not only did my ribs hurt, but my eyes watered and I lost my breath.

It was a group in Kansas where I came across my first sleeper. He was an old man with a Grizzly Adams beard. He had his head down on his chest and snored lightly, his arms folded over his lap. I didn't break my speaking stride at all as I picked out a candy bar and threw a fastball that would have made Nolan Ryan proud. It drilled the guy right in the chest, and he started so hard he nearly fell off his chair. The room went nuts with cheers. His buddies heckled him mercilessly, and every time his eyelids grew a bit heavy, one of them would wing a piece of candy at him. After the meeting, there was a line of people waiting to tell me how impressed they were that Roy stayed awake for most of a meeting for the first time in memory.

In rooms with high ceilings, I'd lob the bars through the air on a high arc. In rooms where the front tables were empty, I would skip or bounce the candy bars off a front table when I wanted one to go to the front rows of attendees. I tried a few skyhooks and other trick shots. People loved the meetings, and compared them to the meetings in previous years, where the presenter had to practically beg for anyone to participate. My meetings, it turned out, were so interactive that there were times I could barely make it through all the slides even though I had built in over a half hour of slack time for questions.

On one of my weeks back at headquarters, my boss called me in to his office and told me that he wasn't sure how it would go over with the big wigs, having a safety nerd throw candy at employees. That being said, he also didn't want me to stop. He had brought an inflatable ball with him to the all-employee meetings he presented. He would chuck it at anyone whose phone rang. He'd been doing it for years, and thought the candy idea was great.

It was close to the end of the meeting cycle, and I was in a challenging room. There was a low ceiling, with even lower wooden cross-members. The group was relatively large, about thirty to forty people. Two other members of the corporate safety department were with me to video record the meeting for employees who missed it and to show upper management how well these meetings were being received; copies of the meeting, burned to DVD, would be sent around the country as needed for those who couldn't attend in person.

My throws had to be low and accurate. It was a breakfast meeting and people had plates of eggs, cups of coffee, and glasses of juice in front of them. Most of the first few throws went really well, but I unintentionally skipped a toss or two off the ceiling. Then the inevitable happened. I threw a bar low to avoid a ceiling skip, and the adjustment ruined my accuracy. I plunked the candy bar off a woman's juice glass. This was one of those incredible events where everything happened in perceived slow motion. I saw the direction the bar was moving. I saw it strike the glass, creating a tiny orange tsunami that leapt over the edge of the glass and soaked the employee's sweater and jeans, the glass tipping over and pouring the rest of the juice in her lap.

The room went silent. I felt the hot redness of embarrassment flush my face red. I was Icarus; I had flown too close to the sun, and now my wings were melted. I took a breath and kept talking, not sure what to do. During the next break I offered an apology, but the look on the woman's face told me it wasn't enough.

When I got back to the office, my boss called me in. Apparently, he said, there had been an odd report to HR about an assault via flying candy bar, and now we had an employee who was requesting to have an entire outfit replaced on the company dime because of juice stains.

"But," I stammered, "orange juice doesn't stain." It was quite possibly the worst response I could have offered. There was no official write-up, and the company didn't take the cost of the employee's new clothes out of my paycheck. I was given strict orders *not* to throw anything anymore, anywhere, any time. In the C-suite, upper management was calling it Snickergate. They had it on video.

I had one more week of meetings to go. The crowd at the first meeting looked up to me expectantly, the candy tossing having grown to legendary status ahead of my arrival. I had a bag of candy bars, and as I spoke, I walked through the crowd, handing them to people. Someone across the room answered a trivia question, raised his hands in the air and said, "Hit me, I'm open!"

Saying "Hit me, I'm open" was like throwing a tennis ball and not expecting a chocolate lab to run after it. I *had* to throw the candy. I wouldn't be able to sleep that night if I didn't. I used a slow, underhand, Granny technique, and the candy bar landed perfectly and safely in the employee's hands. The instructions my boss had issued, something about being careful with my tosses, were lost to my brain (which was now taken over by the twelve-year-old version of myself that wanted to deliver every object, everywhere, every time via throw). Candy bars flew from my personage like I was on a parade float for the remainder of the week.

When I was back at the office the following Monday, I was catching up on emails when my phone rang. The caller ID indicated that it was my boss. I picked up the phone and said good morning. He responded with, "My office," then a click as the phone was hung up.

---

*Nerds behaving badly*

I do not condone blowing off your boss, wife, drill sergeant, etc. My lack of listening skills has been a detriment to me, and typically is not a great method of career or academic advancement. According to my mother, if I had listened to her when I was in junior high, I'd be president right now.

---

I should probably tell you (if you haven't figured it out yet) that getting lectured by a boss, wife, parent, drill sergeant, or other authority figure really isn't new to me. I tend to be what they call "too independent" and oftentimes do things that fall into the realm of "not listening to instructions." I've gotten better as the years have gone, and I really do try to be a well behaved safety nerd as much as I can.

I entered the office and sat down. In a scene reminiscent of Mickey Mouse in *The Sorcerer's Apprentice*, I smiled and sheepishly said, "Good morning, Pete!"

He pointed behind me and sternly said, "Door."

I sighed and closed the door, honestly not sure what it was he was going to be lecturing me about.

"You know," he said, "I usually don't have to tell an employee more than once not to do something. Especially when it's something like a safety professional throwing objects at employees."

I was quick on my toes, and responded with a clever "Oh. Umm. Well, you see…"

"You don't have any more meetings, right?"

I shook my head. "No."

He pointed at me. "Don't do it again. And start listening to me."

I followed half his instructions, anyway.

The next November, I was in front of a group of employees for that year's first meetings. I had no bag of candy bars at my side. When we hit the first trivia question, an employee raised his hand.

Expecting an attempt at an answer for the question, I pointed to him and said, "Go ahead."

"Is it true that you spilled a carafe of cranberry juice on a lady last year?"

The room murmured and a few people said they wanted their candy bars. I explained that it was orange juice, and only a cup of it. The rumor travelled ahead of me as I put on meetings, and now and then an employee with a good sense of humor would tease me by warning that they had a glass of juice and weren't afraid to use it, or by sniping me with a candy bar while I presented. I'm one of the few people who laugh when a random piece of chocolate whizzes past my head. A few groups tried to talk me into throwing candy bars, promising me they wouldn't tell anyone in management. I avoided temptation by not bringing candy or any other things that might be even remotely fun to throw to the meetings.

My lessons from Snickergate reinforced the things I had previously learned about training and other presentations. They are:

1   People generally hate safety meetings and safety training, and with good reason. They are boring.
2   It doesn't take much to make a meeting more interesting if you aren't afraid to think of yourself as a show person rather than just an instructor.
3   Relate to the people you are speaking to and they will engage back to you.
4   Use humor and unexpected surprises to make training a fun experience, and employees will look forward to it.

Don't throw things at people. It isn't nice.

# 27 Winston Churchill said it best

I have no idea what the future definitively holds for the safety profession, I only have my hopes and ideas based on the first leg of my career. Nobody can really read tea leaves or crystal balls; if they could, investing for retirement would be a lot easier. The future of safety will truly be shaped by the people who enter the profession and become its guardians in the years to come.

It is often remarked that the only constant in life is change, but as Gandhi said "You must be the change you wish to see in the world." For safety professionals, this means that not only do we need to operate in a manner that reflects the way we wish to see the safety profession mature, but it means we must take an active role in guiding and molding that change. Without the strong support of professional organizations and robust networking by individuals, our profession cannot advance in a meaningful way.

From 1811 to 1816, a group of English textile workers called "Luddites" protested against technology advances, fearing such advances would reduce the number of jobs available to them. The protesters smashed machines and fought government forces. A mill owner was even assassinated (Wikipedia, n.d.a.). In the end, the Luddites not only failed to stop progress, but history illustrates to us advances in technology not only lead to better lives but to better employment. Technology and labor shift and form around each other through time, and those who embraced technology generally have fared well. It is our challenge in the safety profession to not only embrace technological changes, but to continue learning from our past and present successes and failures. General Patton died because he was not belted into his seat. Does this have any bearing on passenger safety in self-driving cars? Bhopal occurred because the plant design and management systems were poor. How can we ensure that computer modeling is used to sufficiently predict where and how chemical releases could occur? Can we use technology to accurately determine how best to respond to releases? An employee dies because the cover to a floor opening falls through while being set in place. Can architects and engineers receive enough guidance to design other floor openings differently in the future? Scores of people died at The Cocoanut Grove because exits were blocked and exit lighting was unable to be seen and it changed the fire code. Can we learn lessons from fires with smaller death tolls and save more lives?

I don't know the answers to the questions above, but I certainly hold out hope that the safety profession will continue to advance. We don't need more laws in every case, and better use of technology will almost always be more effective than regulations alone.

Winston Churchill said "It is always wise to look ahead, but difficult to look further than you can see." With the rapid progress of technology, the ability to view the future is shortened. Would a safety professional in 1970 predict that less than fifty years later we would have the ability to electronically perform safety audits on a computer smaller than a legal pad?

While technology holds promise for our profession, politics holds uncertainty. Any efforts to dismantle or defund safety and environmental laws must be met with a swift and unrelenting response from professional safety associations. The advances in technology, the new philosophies, research, and learning... It can all be wiped out by the extremist views of a few politicians. Will we allow our profession to continue thriving, growing, and saving lives, or will we allow it to be elastrated, losing all the advances we've made in the last half century?

# Bibliography

American Chiropractic Association. (n.d.). *Back Pain Facts & Statistics*. Retrieved from acatoday.org: www.acatoday.org/level2_css.cfm?T1ID=13&T2ID=68

American Society of Safety Engineers. (n.d.). *What is the American Society of Safety Engineers?* Retrieved from asse.org: www.asse.org/about/history/

Axelrod, A. (2006). *Patton: A Biography*. London: Palgrave Macmillan.

Axelrod, A. (2011). *The Complete Idiot's Guide to the Civil War*. London: ALPHA.

Belsky, J. (2008, September 25). *Rewards are Better than Punishment: Here's Why*. Retrieved from *Psychology Today*: www.psychologytoday.com/blog/family-affair/200809/rewards-are-better-punishment-here-s-why

*BrainyQuote.com*. (n.d.a.). Retrieved from: www.brainyquote.com/quotes/authors/s/sigmund_freud_2.html

*BrainyQuote.com*. (n.d.b.). Retrieved from: www.brainyquote.com/quotes/quotes/w/wedwardsd377112.html

Bramble, D., & Lieberman, D. (2004). Endurance Running and the Evolution of Homo. *Nature*, 432, 345–352.

Brandt, N. (2003). *Chicago Deathtrap: The Iroquois Theater Fire of 1903*. Carbondale, IL: Southern Illinois University Press.

Brennan, C. (2013). *The Bite in the Apple: A Memoir of my Life with Steve Jobs*. New York: St. Martin's Press.

Cambridge Center for Behavior Studies. (n.d.). *Introduction to Behavioral Safety*. Retrieved from http://www.behavior.org/resource.php?id=330

Carrillo, R. A. and Samuels, N. (2015). Safety Conversations: Catching Drift & Weak Signals. *Professional Safety*, 60(1), 22–32.

Casselman, B. (2010, May 20). *Rig Owner had Rising Tally of Accidents*. Retrieved from *Wall Street Journal Online*: www.wsj.com/articles/SB10001424052748704307804575234471807539054

Centers for Disease Control. (n.d.). *Overweight & Obesity*. Retrieved from www.cdc.gov/obesity/

Chances for the Inventor. (1902, September 9). *The Friend*, vol. 76, p. 28.

Chernow, R. (2005). *Alexander Hamilton*. New York: Penguin Books.

Cherry, K. (n.d.). *What is a conditioned response?* Retrieved from about.com: http://psychology.about.com/od/cindex/g/condresp.htm

*civilwarhome.com*. (n.d.). Retrieved from Jackson's Confederate Military History Biography: www.civilwarhome.com/CMHjacksonbio.html

Coleman, P. (2010). *A Woman Unafraid: The Achievements of Francis Perkins*. Bloomington, IN: iUniverse.

D'Este, C. (1995). *Patton: A Genius for War.* New York: Harper Collins.

Eckerman, I. (2005). *The Bhopal Saga: Causes and Consequences of the World's Largest Industrial Disaster.* Hyderabad, India: Universities Press.

Ehrman, M. (2006). *Take a Nap! Change Your Life.* New York: Workman Publishing Company.

Emsley, J. (2011). *Nature's Building Blocks: An A–Z Guide to the Elements.* New York: Oxford University Press.

Esposito, J. (2006). *Fire in the Grove: The Cocoanut Grove Tragedy and its Aftermath.* Cambridge, MA: De Capo Press.

Ferdman, R. (2014, September 25). *The Pay Gap Between CEOs and Workers is much Worse than you Realize.* Retrieved from *The Washington Post*: www.washingtonpost.com/news/wonkblog/wp/2014/09/25/the-pay-gap-between-ceos-and-workers-is-much-worse-than-you-realize/

Fernandez, M. (2014, April 22). *Lax Oversight Cited as Factor in Deadly Blast at Texas Plant.* Retrieved from *The New York Times:* www.nytimes.com/2014/04/23/us/lack-of-oversight-and-regulations-blamed-in-texas-chemical-explosion.html?_r=0

Fernandez, M., & Schwartz, J. (2013, April 18). *Plant Explosion Tears at the Heart of a Texas Town.* Retrieved from *The New York Times*: www.nytimes.com/2013/04/19/us/huge-blast-at-texas-fertilizer-plant.html?pagewanted=all

Fisher, T. (2011, December 6). *How Often do Men and Women Think About Sex?* Retrieved from *Psychology Today:* www.psychologytoday.com/blog/the-sexual-continuum/201112/how-often-do-men-and-women-think-about-sex

Frank, Thomas (n.d.). *Thomas Frank Quotes.* Retrieved from BrainyQuote.com: www.brainyquote.com/quotes/quotes/t/thomasfran484710.html

Gablet, N. (2006). *Walt Disney: The Triumph of the American Imagination.* New York: Random House.

Goldstein, B. (1997). Saving the Phenomena: The Background to Ptolemy's Planetary Theory. *Journal for the History of Astronomy*, 28(1), 1–12.

Greene, N. (n.d.). *Ptolemy Biography.* Retrieved from about education: http://space.about.com/cs/astronomerbios/a/ptolemybio.htm

*Grimshaw v. Ford Motor Company*, 119 Cal. App. 3d 757 (California 1981).

Harris, S. (1994). *The United States Pony Club Manual of Horsemanship: Basics for Beginners – D Level.* New York: Howell Book House.

Heinrich, H. (1931). *Industrial Accident Prevention: A Scientific Approach.* New York: McGraw-Hill.

*Helmet Related Statistics from Many Sources.* (n.d.). Retrieved from helmets.org: www.helmets.org/stats.htm

Henslin, J. (2008). *Sociology: A Down to Earth Approach.* London: Pearson Education.

Hopkin, M. (2005). *Ethiopia is Top Choice for Cradle of* Homo Sapiens. Retrieved from Nature News: www.nature.com/news/2005/050214/full/news050214-10.html

Jessen, C. (2000). *Temperature Regulation in Humans and Other Mammals.* Berlin: Springer.

Kenney, C. (1999, May). *Did a Mystery Gas Fuel the Cocoanut Grove Fire?* Retrieved from Firehouse: www.firehouse.com/article/10544181/did-a-mystery-gas-fuel-the-cocoanut-grove-fire

Krumm, D. (2000). *Psychology at Work: An Introduction to Industrial/Organizational Psychology.* New York: Macmillan.

LifeSpan. (n.d.). *Posture Perfect: Standing Up for Your Back.* Retrieved from www.lifespan.org/conditions-treatments/wellness-library/posture-perfect-standing-up-for-your-back

Livestrong. (2015, April 14). *How Many Calories do you Burn Standing all Day?* Retrieved from www.livestrong.com/article/312386-how-many-calories-do-you-burn-standing-all-day/

Lyons, L. (2006, April 3). *Why do Cats Purr?* Retrieved from *Scientific American*: www.scientificamerican.com/article/why-do-cats-purr/

MacLaury, J. (n.d.). *The Job Safety Law of 1970: Its Passage was Perilous.* Retrieved from Department of Labor: www.dol.gov/dol/aboutdol/history/osha.htm

Manuele, F. (2002). *Heinrich Revisited: Truisms or Myths.* Itasca, IL: National Safety Council.

Maraniss, D. (1999). *When Pride Still Mattered, A Life of Vince Lombardi.* New York: Simon & Schuster.

Meachum, J. (2013). *Thomas Jefferson: The Art of Power.* New York: Random House Trade Paperbacks.

Merriam, S., & Brockett, R. (2007). *The Profession and Practice of Adult Education: An Introduction.* San Francisco, CA: Jossey-Bass.

Metcalf, R. (2000). *Ulmann's Encyclopedia of Industrial Chemistry.* Weinheim, Germany: Wiley-VCH.

Millard, C. (2009). *The River of Doubt: Theodore Roosevelt's Darkest Journey.* New York: Random House.

Mullis, S. (2012, July 16). *What are the Advantages and Disadvantages of Standing Desks?* Retrieved from *Forbes.com*: www.forbes.com/sites/quora/2012/07/16/what-are-the-advantages-and-disadvantages-of-standing-desks/

*New York Times* (2015) Surviving the Nazis, Only to be Jailed by America. February 7.

OSHA. (1993). *Ergonomics Program Management Guidelines for Meatpacking Plants.* Retrieved from www.osha.gov/Publications/OSHA3123/3123.html

OSHA. (2012, January). *Injury and Illness Prevention Programs.* Retrieved from www.osha.gov/dsg/InjuryIllnessPreventionProgramsWhitePaper.html

OSHA. (n.d.a.). *OSHA FactSheet: OSHA Inspections.* Retrieved from www.osha.gov/OshDoc/data_General_Facts/factsheet-inspections.pdf

OSHA (n.d.b.). *OSHA Injury and Illness Recordkeeping and Reporting Requirements.* Retrieved from www.osha.gov/recordkeeping/index.html

Partch, H. (1979). *Genesis of a Music.* Boston, MA: Da Capo Press.

Pontius, J., et. al. (n.d.). *Initial Sequence and Comparative Analysis of the Cat Genome.* Retrieved from Genome Research: http://genome.cshlp.org/content/17/11/1675.full

Prince, K. (2013, February 5). *The Difference Between Positive/Negative Reinforcement and Positive/Negative Punishment.* Retrieved from bcotb.com: http://bcotb.com/the-difference-between-positivenegative-reinforcement-and-positivenegative-punishment/

Rogers Commission. (1986). *Report of the Presidential Commission on the Space Shuttle Challenger Accident.*

Rowe, M. (n.d.). *Barbershop Harmony Society.* Retrieved from barbershop.org: http://barbershop.org/news-a-events-main/46-mike-rowe-interview.html

Roy, W., Campbell, J., & Toure McDiarmid, T. (1999). *Snake Species of the World: A Taxonomic and Geographic Reference.* Washington, DC: Herpetologists' League.

Salvendy, G. (2012). *Handbook of Human Factors and Ergonomics.* Weinheim, Germany: Wiley.

Sass, S. (1989, Spring). A Patently False Patent Myth. *Skeptical Inquirer*, 13, 310–313.

Schaefer, J. (2014, April 15). *The Root Causes of Low Employee Morale.* Retrieved from American Management Association: www.amanet.org/training/articles/The-Root-Causes-of-Low-Employee-Morale.aspx

Schrenk, F., Kullmer, O., & Bromage, T. (2007). The Earliest Putative Homo Fossils. In *Handbook of Paleoanthropology* (pp. 1161–1631) Henke, W. and Tatersall, I. Berlin Heidelberg: Springer-Verlag.

Smith, M. (1996). *Ptolemy's Theory of Visual Perception- An English Translation of the Optics.* Philadephia, PA: The American Philosophical Society.

Social Welfare History Project. (n.d.). *Child Labor Reform.* Retrieved from www.socialwelfarehistory.com/programs/child-labor-reform-an-introduction/

Spine-health. (n.d.). *Avoid Back Injury with the Right Lifting Techniques.* Retrieved from www.spine-health.com/conditions/sports-and-spine-injuries/avoid-back-injury-right-lifting-techniques

State of California. (n.d.). *Injury and Illness Prevention Etool.* Retrieved from ca.gov: www.dir.ca.gov/dosh/etools/09-031/cost.htm

Stein, Leon (1962). *The Triangle Fire.* Ithaca, NY: Cornell University Press.

Swisher, C., Curtis, G., & Lewin, R. (2001). *Java Man: How Two Geologists Changed our Understanding of Human Evolution.* Chicago, IL: University of Chicago Press.

Tinsley, A., & Ansell, R. (2011, October 13). *Bhopal's Never Ending Disaster.* Retrieved from *The Environmentalist*: www.environmentalistonline.com/article/2011-10-13/bhopal-s-never-ending-disaster

Twenge, J. (2006). *Generation Me.* New York: Free Press.

U.S. Coast Guard. (2011, September 14) *Report of Investigation into the Circumstances Surrounding the Explosion, Fire, Sinking and Loss of Eleven Crew Members Aboard the Mobile Offshore Drilling Unit Deepwater Horizon in the Gulf of Mexico, April 20–22, 2010.* Washington, DC: Bureau of Ocean Energy Management, Regulation, and Enforcement, United States Coast Guard.

U.S. Department of Transportation Highway Traffic Safety Administration. (1968). *Title 49 of the United States Code, Chapter 301, Motor Vehicle Safety Standards No. 208.*

U.S. Government Printing Office. (2011). *Investigating OSHA's Regulatory Agency and its Impact on Job Creation.* Washington, DC: U.S. Government Printing Office.

Valdes-Dapena, P., & Yellin, T. (n.d.). *GM: Steps to a Recall Nightmare.* Retrieved from CNN Money: http://money.cnn.com/infographic/pf/autos/gm-recall-timeline/

Varma, R., & Varma, D. (2005). The Bhopal Disaster of 1984. *Bulletin of Science, Technology and Society,* 25(1), 37–45.

Vlasic, B., & Apuzzo, M. (2014, March 19). Toyota is Fined $1.2 Billion for Concealing Safety Defects. *New York Times.*

*Which U.S. Presidents Owned Slaves?* (n.d.). Retrieved from http://pres-slaves.zohosites.com/

Wikipedia. (n.d.a). Luddite. Retrieved from wikipedia: https://en.wikipedia.org/wiki/Luddite

Wikipedia. (n.d.b). M113 Armored Personnel Carrier. Retrieved from wikipedia: https://en.wikipedia.org/wiki/M113_armored_personnel_carrier

Wollheim, R. (1981). *Sigmund Freud.* Cambridge: Cambridge University Press.

Womack, J. (2011). *Gemba Walks.* Cambridge, MA: Lean Enterprise Institute.

Work, D. (2012). *Lincoln's Political Generals.* Urbana, IL: University of Illinois Press.

Yeager, C. and Janos, L. (1985). *Yeager: An Autobiography.* New York: Bantam Books.

Zaloga, S. (2010). *George S. Patton: Leadership, Strategy, Conflict.* Oxford: Osprey Publishing.

Zimmer, C. (2011, January). *100 Trillion Connections: New Efforts Probe and Map the Brain's Detailed Architecture.* Retrieved from *Scientific American*: www.scientificamerican.com/article/100-trillion-connections/

# Index

accidents 34–5, 128–9; behavior as contributing factor for 23–4; blame for 23–4; chemical 63–6, 128–9, 135; cleanup costs 74; downtime 74; driving 80–2; fault versus fact finding 63; FEMA 93; interviewing witnesses 94; lack of common sense and 32; poor decisions and 133; preventable 16; pyramid 23. *See also* fatalities

American National Standards Institute (ANSI) 115

American Society of Safety Engineers (ASSE) 45, 67

ANFO 65–6

ANSI. *See* American National Standards Institute

APC. *See* armored personnel carrier

armored personnel carrier (APC) 80–2

ASSE. *See* American Society of Safety Engineers

automobile industry: accidents 16, 25

Baby Boomers 136, 137

barriers 60–1

BBS. *See* behavior-based safety

behavior: bad 141–2; changes in 41–2; as contributing factor to accidents 23–4; fearless 1; natural 37; relationships 48, 59; rewards and 42–3, 108–9, 132; unsafe 132. *See also* behavior-based safety

behavior-based safety (BBS) 23, 36–43, 108. *See also* behavior

bicycle helmets 33

BLS. *See* Bureau of Labor Statistics

Board of Certified Safety Professionals 8

Bureau of Labor Standards 120, 124

Bureau of Labor Statistics (BLS) 5, 49

Candhi 143

catwalk 114

CDC. *See* Centers for Disease Control

Centers for Disease Control (CDC) 20

Certified Safety Professional (CSP) 8

CFR. *See* Code of Federal Regulations

*Challenger* 24–5

Chamber of Commerce 122

chemical accidents 63–6, 128–9, 135

children: workplace safety and 118

Churchill, Winston 143–4

circular openings 96

citation: contesting 78; penalties 79

claims manager 100

Cocoanut Grove fire 128–9

Code of Federal Regulations (CFR) 5

collaboration 110

common sense: description of 30–5; lack of 32

communication 108

competition 57

control-reliable 110

costs: cleanup 74; hidden 73; overtime 74

creativity 48

cryogenic cylinders 27

CSP. *See* Certified Safety Professional

C-suite 70, 77

culture: changes in 99; references in the 21st century 67; safety and 5

DAF (dumb ass factor) 130–3

danger 49–50

debriefing 113

decisions: accidents and 133

Dekker, Sidney 58

Deming, W. Edwards 12, 48; principles 58–62; red bead experiment 54

discipline 114